THE AIR FORCE AND THE GREAT ENGINE WAR

THE AIR FORCE AND THE GREAT ENGINE WAR

ROBERT W. DREWES

1987

NATIONAL DEFENSE UNIVERSITY PRESS

WASHINGTON, DC

National Defense University Press Publications. To increase general knowledge and inform discussion, NDU Press publishes books on subjects relating to US national security. Each year, in this effort, the National Defense University, through the Institute for National Strategic Studies, hosts about two dozen Senior Fellows who engage in original research on national security issues. NDU Press publishes the best of this research.

In addition, the Press publishes especially timely or distinguished writing on national security from authors outside the University, new editions of out-of-print defense classics, and book based on conferences concerning national security affairs.

Proofread by William A. Palmer, Jr., Cheltenham, Maryland, under contract DAHC32–87–A–0015. Editorial Experts, Inc., Alexandria, Virginia, indexed this book under contract DAHC32–86–A–0009.

NDU Press publications are sold by the US Government Printing Office. For ordering information, call (202) 783-3238, or write to: Superintendent of Documents, US Government Printing Office, Washington, DC 20402.

First printing, March 1987.

Library of Congress Cataloging-in-Publication Data

Drewes, Robert W.
The Air Force and the great engine war.

Bibliography: p.
1. United States. Air Force—Procurement—History. 2. Airplanes, Military—Turbofan engines. 3. Pratt & Whitney Aircraft Group—History. 4. General Electric Company—History. I. Title.
UG1123.D74 1985 358.4'182 86–12636

CONTENTS

FOREWORD

HIGHLY PUBLICIZED ACCOUNTS of abuse in military weapons procurement have raised both citizen awareness of and citizen concern with the properly monitored spending of US defense dollars. Not long ago, media reports of spare parts overpricing and related problems ignited harsh public criticism of the handling of the multibillion dollar defense contracts for the F100 jet engine. According to Colonel Robert Drewes, US Air Force, though, the outcome of the subsequent "Great Engine War" calls not for criticism, but for praise for the Department of Defense.

Long before the public became aware of the controversy, the Air Force was grappling with the problems of the F100 high performance engine and the contract for its procurement and maintenance. As difficulties mounted in negotiations with the sole-source supplier, the Air Force, Navy, and Congress held their ground and eventually prevailed. The account of their combined efforts is an encouraging story about the Department of Defense and the US Government "setting things right," a story that has not been fully told before.

The case is not closed on jet engine contracting, or any other kind of defense contracting, but the Great Engine War is welcome reassurance that US defense dollars—closely monitored—will be spent wisely.

Bradley C. Hosmer
Lieutenant General, US Air Force
President, National Defense University

PREFACE

I WROTE THIS BOOK because the telling of these events is overdue. In the midst of barrages of criticism of how the military does business, someone had to tell a different story, a story in which, at least for the moment, the Air Force can take pride. Whether the pride can endure depends on how well additional gains are made with the opportunities now available.

The story focuses on the complex considerations and interactions which permeate every step in bringing a major weapon system from a mental image to the drawing board, to production, and on to operational use. Along the way the work is saturated with incessant concerns over congressional actions, inter-Service rivalry, interstaff squabbles, and contractor posturing as well as hard-core problems with the technology. The objective of this book is to demonstrate how individuals working in an environment of seemingly endless distractions and frustrations can still have a vision of what makes sense and persist in making the ideal a reality.

Initially this book was to concentrate only on the events beginning in the early 1980s when the Air Force started bringing General Electric and Pratt and Whitney together again in a head-to-head competition for future requirements. However, as research progressed it became clear that to appreciate the significance of this formal competition, the earlier struggles with the engine and mounting emotional intensity must be understood as well. Furthermore, the message must also be conveyed that the problem engine, the F100 built by Pratt, is, nevertheless, a marvelous machine. Gene Bryant, the Air Force F100 program manager, explains that "people have asked a lot of the engine and it has given a lot."[1] General Bellis, who managed the concurrent development of the F–15 fighter and the F100 engine, states, "Over the 15 years since we started the F100 development program, it still is the highest performing engine in the world with corresponding fuel efficiencies. It has a better operational record in the USAF inventory than any other fighter engine."[2] Therefore, in telling the story I have tried to include the many positive aspects of the engine's performance.

Likewise, this story is not intended to castigate a particular contractor. Specific problems arose, individuals reacted in certain ways, miscommunications occurred, and in retrospect, it seems easy to see how situations could have been handled better. But, to identify these problems is not to categorically denounce a contractor or a project. Certainly, such is not my intention. Events occurring in times past do not necessarily portend how individuals or institutions will face future challenges.

I wish to gratefully recognize Major General Bernard Weiss for his suggestions in launching the research in this topic. Special recognition is owed as well to Dean Gissendanner and Tack Nix at the Pentagon, Ron Mutzelburg at Wright-Patterson Air Force Base, and Nick Constantine, enjoying retirement in Alexandria, Virginia, for their always thoughtful insights and assistance with this story. In addition, the comments from Generals Bellis and Slay were especially valued. Furthermore, at a key juncture in my research, Professor Herm Stekler of the Industrial College of the Armed Forces faculty wisely suggested additional productive sources for background information on the contractors.

Most important, all of my hundreds of hours of interviews and examinations of documents would have been wasted without the professional assistance of the National Defense University Research and Publication Directorate.

R.W.D.

THE AIR FORCE AND THE GREAT ENGINE WAR

INTRODUCTION

THIS BOOK examines the Great Engine War, referred to by Secretary of the Air Force Verne Orr in 1984 as "perhaps the most significant Air Force acquisition initiative in the past decade."[1] The "war" pitted Pratt and Whitney Aircraft (part of the United Technologies conglomerate) against General Electric in head-to-head competition for multibillion dollar defense contracts to provide high performance engines for the nation's frontline fighter aircraft. Throughout the war, from early skirmishes as each firm maneuvered into position, through periods of behind-the-scenes planning, to the outcome of the official competition in 1984, the Air Force dueled constantly with the contractors and ultimately set the rules of engagement for the biggest battle of all, thus far.

More than 10 years after the Air Force decided Pratt had won the competition against General Electric to develop and manufacture the F100 engine for the F–X, the Air Force changed its mind and reopened engine competition. After receiving thousands of engines from Pratt, the Air Force set out to find a better deal for the remaining billions of dollars of engines still required. Why?

Historically, after a defense contractor had competed and won a contract to design, develop, and produce any portion of a major weapon system (such as the engine), he was virtually assured (through operation of the marketplace) of receiving all follow-on business for the product without competing again. Determination by economics usually guarantees this sole-source, noncompetitive position. The costs incurred to design a competitive product, to set up a production line, to provide an experienced labor force, and to establish the field service that supports operational use of a product usually preclude a profit-motivated business from submitting a new competitive proposal.

How can an experienced, incumbent contractor in a capital-intensive industry ever lose follow-on business? Won't the incumbent always have a lower cost base than anyone else trying to get started with a competitive product? Over decades of procurement experience this uncomplicated assessment has proven accurate, yet in the case of the F100 engine, it did not hold true. General Electric, an anxious, skillful competitor, sought an opportunity to break precedent by challenging Pratt's position. The special circumstances which facilitated GE's quest and the consequences of the GE victory are the subjects of this book.

Most of those who were close to the last battle of the Great Engine War, which led to Secretary Orr's decision in February 1984 to split future engine sales between Pratt and GE, view the engine competition as one of the finest accomplishments of their careers. For instance, General Lawrence Skantze, commander of Air Force Systems Command, the Air Force's primary research, development, and acquisition organization, speaks of this engine competition as "the most exciting thing I've ever been associated with in the acquisition business."[2] Secretary Orr highlighted the advantages of competition in a news release, stating, "We set the mark very high and we met our mark."[3] Others however, consider the competition, specifically the way the competition was conducted, unprofessional bordering on unethical. Such critics, although distinctly in the minority, believe the engine competition has unjustifiably jeopardized combat effectiveness and pilot survivability. This study attempts to cut through the strong, sometimes emotional opinions of the "combatants" and then to discern and report reality.

The principal objective of this examination is to identify the lessons learned. An Air Force general stated, "History impresses people more than philosophy. I'll tell you something. The next time the Air Force tells [contractors] to pay attention to something, they will. That's what this did. It got their attention."[4] However, having focused industry and Government concentration on this unprecedented change in acquisition, the Air Force must be careful to draw right conclusions. If the competition was successful, what conditions made it so? Can the Air Force reproduce those conditions to reap similar benefits in other major military procurements? And if the competition was a failure, can the Air Force salvage for the future something beneficial from the experience?

An underlying theme of this examination is that although the "Great Engine War" is a label usually given to the GE versus Pratt competition of the early 1980s, the "war" actually began much earlier. Perhaps the contractors did not fully realize it at the time, but by their decisions in the early 1970s they were positioning themselves for the full-scale battles of a decade later. Similarly, the Air Force and, to a somewhat lesser but still pivotal degree, Congress and the Navy highly influenced the circumstances of the official competition in 1983 by their past actions. In order to capture the whole story, the book begins by returning to the late 1960s and examining in detail the environment in which the F100 was developed for the F–X, later designated the F–15 after the McDonnell Douglas design was selected for the aircraft. After depicting the major events in those early years, the book describes the results of operational use of the engine when severe problems were encountered and solutions were found only with great difficulty. Problem after problem arose. Moreover, in the middle of these technical matters lay a hard core series of Government-to-contractor *personality* issues.

Fundamental to understanding these historical events is appreciation for the management styles, business expectations, and communication skills of the key participants, factors which highly influenced the head-to-head competition. As the story unfolds, the reader will see that the major lessons from the Great Engine War do not concern the specific techniques or procurement methodologies employed by the Government or the contractors. The primary value in studying this unique acquisition is in understanding the people who were involved, appreciating the pressures under which they made decisions, their adaptability to change, their determination to succeed, and the key role of chance in the outcome of their efforts.

1
GETTING OFF THE GROUND

On the previous page, the F–15 Eagle. When the role of the F–X (later named the F–15) became air-to-air performance, the Air Force was determined to build its own design. (An official US Air Force photo, T. Sgt. Rod Prouty, photographer.

THE MID-1960S WERE UNSETTLED TIMES marked by increasing concern for the tactical air power of the United States Air Force, particularly its ability to control the skies in war. Disturbing evidence appeared to indicate the Soviets had narrowed the gap of performance capabilities between our and their frontline fighters. Most disturbing of all, the evidence suggested almost certain Soviet air superiority in the mid-1970s and beyond—unless, of course, the USAF could develop and deploy a counter to the anticipated Soviet threat.

THE THREAT

One of the first pieces of evidence emerged from statistics on our performance in air-to-air combat over North Vietnam. Our success ratio was about 20 percent of the effectiveness in the Korean war. Over North Vietnam the United States was losing one airplane for every 2.5 enemy aircraft destroyed. Nearly 25 years before in North Korea, the ratio had been one US loss for every 12 enemy.[1] As early as April 1965, for example, we saw evidence of the shift in air superiority when supposedly obsolete MiG–15s shot down two of our F–105s over North Vietnam. True, the F–105s weren't fully maneuverable since they still carried their bomb loads at the time, but the incident showed the United States performing below expectations in air-to-air combat. Then in July 1967 came a real shocker. Although the Soviets had not mounted an air show or Aviation Day for six years, in July 1967, at the new Domodedovo Civil Airport south of Moscow, the Soviets surprised the world by unveiling 12 new *and* advanced military aircraft, including two, the Flogger and Foxbat, clearly superior to any fighter the United States had either in inventory or approved for new development and acquisition.[2]

Today the MiG–23, Flogger, a single seat, swing-wing air combat fighter of Mach 2 speed, forms the backbone of Soviet Frontal Aviation tactical air forces.[3] The MiG–25, Foxbat, is even faster since it was first designed to intercept the high-altitude Mach 3 B–70. Though President Kennedy canceled the B–70, except for some continuing research work, the Soviets, impressed by the MiG–25, pressed ahead. When officially unveiled, these new Soviet fighters and their 10 other new stablemates generated considerable publicity in the United States and made enduring impressions on USAF leadership.[4] Members of the House Appropriations Committee noticed them also, citing the

Foxbat, in particular, as evidence the United States had fallen behind the Soviets in military aircraft development.[5]

INTERNAL DEBATES

While the Soviets had been building new, powerful air superiority fighters, the US military planned

The MiG–23 (below) and MiG–25 (bottom) surprised the House Appropriations Committee.

and produced paperwork exercises and discussions, but no hardware. Our involvement in Southeast Asia was hamstringing the USAF in several ways. Because the conflict had grown very expensive, the US economy and the Department of Defense budget could ill afford a new, expensive fighter development program.[6] Furthermore, the mid-1960s belonged to the McNamara era of exhaustive quantitative analyses and cost-benefit tradeoffs, leaving the Air Force faced with depressing demands to justify every aspect of the requirement for a new fighter.

Even the primary mission of the Air Force's newest proposed aircraft, tentatively called the F–X until selection of a specific design, was under "analysis." The idea of the F–X had been formalized in January 1965. Air Force headquarters and the Tactical Air Command wanted a new fighter. However, some advocates wanted a multipurpose replacement for the F–4 while others insisted on a fighter dedicated solely to the air-to-air combat mission.[7] Multipurpose capability, combining close air support and air superiority, offered greater cost effectiveness in the minds of some analysts. Instead of buying two fighters, one for each mission, buy an aircraft capable of both missions.

Conversely, in the mid-1960s the Tactical Air Forces (Tactical Air Command, United States Air Force, Europe, and Pacific Air Force) were emerging as a united assertive arm of the Air Force during annual struggles for slices of the overall defense budget. In the relatively short history of the USAF, the Strategic Air Command's concern for a superior strategic deterrent had dominated the tussle for dollars. The Vietnam war, however, brought into sharp public focus the need for weaponry improvements. For the Tactical Air Forces, these improvements included an F–X design solely for air superiority to counter the Soviet threat.[8]

As the debates over cost and aircraft design alternatives continued, a long-standing rivalry of high emotional intensity began to play a daily role in the bureaucratic gestation of the F–X. The Navy also needed a new fighter to counter the Soviet threat to the fleet. Could a single design satisfy both Air Force and Navy needs? If so, which Service should have the lead in selecting the design and managing the development effort? A lot of emotional baggage traveled with the answer to these questions. The Air Force still resented having the Navy F–4 forced upon it by Secretary of Defense Robert McNamara rather than having approval to build its own design.

The Navy, for its part, unenthusiastically embraced McNamara's decision to develop a bi-Service F–111 under Air Force lead-management responsibility. Thus, the Air Force was determined never again to be forced into buying an airplane initially designed by the Navy, and the Navy was determined to avoid another joint program like the F–111. These ill feelings, aggressions, and ambitions provided fertile soil for the seeds of a long fight. Defense planners had to settle the primary role for the F–X without ignoring volatile matters of inter-Service rivalry.

How long might these debates have remained unsettled, helplessly entwined in still more studies, meetings and reports, without the stimulus for real action provided by the evidence of Soviet progress? Slowly the primary role of the F–X became air-to-air performance. General John P. McConnell, Air Force Chief of Staff, in congressional testimony in February 1968 said, "We have a compelling need to initiate a program to develop a modern fighter aircraft like the F–X with superior air-to-air performance and substantial *air-to-ground capability* [emphasis added]."[9] However, just six months later General McConnell told the Senate Preparedness Subcommittee:

> The appearance of the Mach 2.8 Mikoyan Foxbat interceptor and other new Soviet fighters ''dictates that we produce a fighter aircraft *optimized for the air-to-air role* [emphasis added].... To keep pace with the Soviet advances in this area our own fighter force must be modernized by the mid-seventies.''[10]

The Air Force helped itself when its senior leadership agreed to buy the A–7, another plane developed by the Navy. Anticipating bureaucratic jousting with the Office of the Secretary of Defense (OSD) to sell the F–X air-to-air role, the Air Force agreed to buy some A–7Ds for the air-to-ground mission. By demonstrating flexibility and willingness to cooperate, the Air Force positioned itself to argue later that, whenever possible, it had accepted existing technology regardless of which Service had developed it. However, the Air Force would argue, because the air-to-air mission requirement remained unsatisfied, the Air Force needed OSD approval to move quickly to acquire a superior, usable aircraft, to get ''rubber on the ramp'' by the mid-1970s. In this manner the issue of the F–X primary role was settled and likewise, but more grudgingly, the notion of another bi-Service fighter went away for a time as well.

However, in abandoning the idea for a common aircraft, the Director of Defense Research and Engineering (DDR&E) prevailed. He recommended that the Air Force and Navy conduct a joint engine development program. Although it might be argued that the missions of the Air Force and Navy fighters were dissimilar enough to require independent designs, their propulsion requirements were very much the same. Even though the administrative mechanics of how the Services would work together to conduct the program were not stated, at least at first, the notion of a common engine, rather than a common aircraft, was palatable. The Air Force subsequently won approval to develop and build its own air superiority fighter by using an engine jointly developed with the Navy.

The engine was probably the single most important component of the aircraft. The key to countering the Soviet threat would be superior F–X maneuverability, and the key determinant of that maneuverability would be engine performance. An excess of maneuvering power would provide the F–X pilot with additional thrust to close on the enemy, to evade, and to climb high quickly.[11]

Usually engine performance is measured by pounds of thrust produced relative to the weight of the engine. Accordingly, the thrust-to-weight ratio is a common standard for evaluating the power and complexity of engines. Until the F–X came along, engines with a 4-to-1 thrust-to-weight ratio were the best that could be built. However, based on US analyses of the threat, clearly more was needed. Early requirements statements set the propulsion objective for the F–X engine at a 10-to-1 ratio; later ones reduced that ratio to 8-to-1. The F–X would be powered by two such engines. With this much power, the Air Force would have a racehorse capable of vertical climbs, a spectacular demonstration of engine power. Such engines, in concert with the geometry of the airframe, would provide a weapon of superior maneuverability and lethality.

A modern jet engine, such as that needed to power the F–X, is an extremely complicated machine with thousands of precisely connected parts, all of which must function perfectly. Basically, the engine pulls in air through the front and then compresses it through a series of compressor stages to very high pressures. The compressor stages are fans, pointed in the same direction and with power enough to compress air about 23 times greater than the pressure at which it enters the engine. Once compressed, the air combines with fuel, ignites explosively, and propels out the rear of the engine to provide the thrust. As the high-pressure exhaust leaves the engine, it passes through a series of turbine blades. The force of the exhaust turns the blades, much like the blades of a windmill. The energy from the rotating turbine is rechanneled forward to the compressor stages to keep the engine running.[12]

The stresses on the engine parts are almost literally beyond comprehension. A technical marvel was needed for the F–X. Unfortunately, according to Senator Barry Goldwater, a former military pilot and a retired major general in the Air Force Reserve, "between 1960 and 1970 engine development in the country practically stood still."[13] Almost no effort was made to increase thrust-to-weight ratios or improve fuel consumption.

To satisfy new needs, therefore, giant strides had to be made quickly to cope with the dynamics at work to yield an engine powerful enough, for example, to propel the F–X up to two miles in altitude in 10 seconds. Engineers would have to understand the problems and solutions in designing turbine blade tips to withstand almost supersonic speeds of turbine rotation and to tolerate powerful friction forces between rapidly moving metal parts. Another tough engineering problem was how to keep a steady air flow into the engine while the aircraft made rapid, high-G evasive maneuvers in air-to-air combat. Add the limitation of the engine's weight allowance imposed to keep the thrust-to-weight ratio high, and one begins to appreciate just a few of the innumerable technical challenges. It was an enormous undertaking. More than any other single factor, the ability to develop, manufacture, and operate the engines to propel the aircraft would determine the success of the F–X program and the US ability to meet the Soviet threat by the mid 1970s.

PRESSURE TO SUCCEED

From the outset, the program to develop and produce the F–15 aircraft was under intense pressure to succeed. This pressure forced some decisions to be made earlier, tests to be completed sooner or omitted, and engines to be placed in operation faster than would ideally have been the case. First, by the mid-1970s, to counter the formidable

Major engine components of a modern jet engine. (Courtesy of US Air Force.)

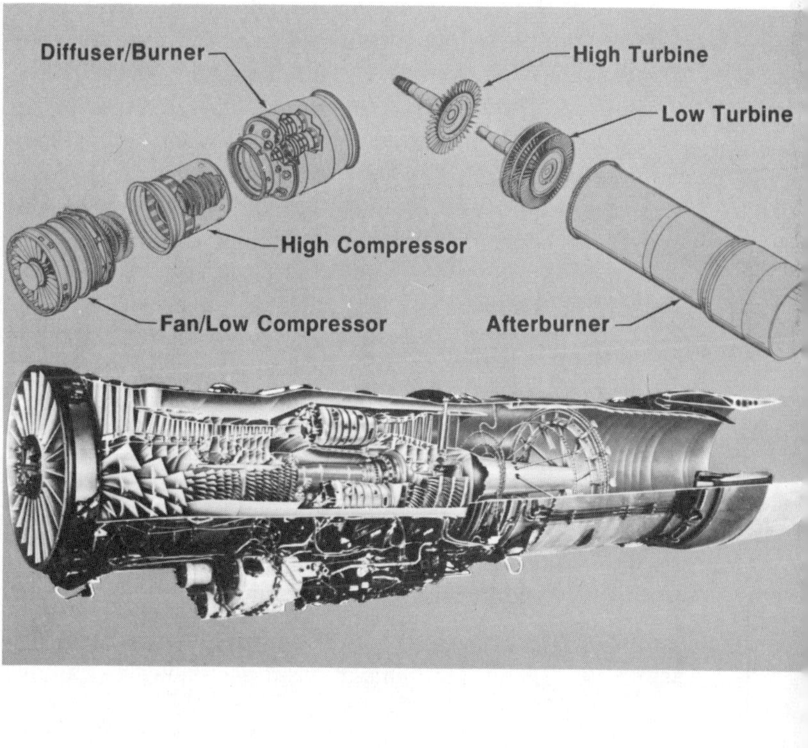

Soviet threat, the Flogger and Foxbat, the United States needed operational, deployed F–15s ready to engage these Soviet aircraft in battle, if necessary. Second, the Air Force needed to prove to the Office of the Secretary of Defense, to Congress, and to the public that it could manage a major development program, successfully meeting all technical, schedule, and cost objectives. Fresh from the painful public outcry from cost overruns, delays, and performance deficiencies in producing the giant C–5A airlifter, the Air Force was especially keen on re-establishing public trust in Air Force business skills through the F–15 program.

Similarly, the Office of the Secretary of Defense was determined to show through successes of the F–15 program that the more prudent management practices of the Nixon administration, sworn into office in January 1969, could achieve the positive results expected by the public. In pursuit of better procedures, Deputy Secretary of Defense David Packard had instituted certain management principles and practices to correct mistakes of the past. Among the improvements were the ''demonstration milestone concept'' and ''streamlining'' the management structure.

The F–15 would be the first Air Force production program to operate under the demonstration milestone concept of holding off a commitment to produce in volume until the technology critical to further development had been satisfactorily demonstrated during the development phase. Unlike earlier programs in which the Government had little visibility of actual technical accomplishments during much of the contractors' work, with this new concept the Government would be assured that visibility.

The overall program was divided into a series of discrete time spans, the completion of each span measurable by observing a demonstrated accomplishment, such as completion of a specific test. Only when that completion, or milestone, had been demonstrated would funds be released to proceed working toward the next milestone. By proceeding in this manner, the Air Force would be protected from moving too quickly into production, incorporating technology which perhaps had not been sufficiently developed and harnessed for the intended purpose in the weapon. Too frequently in the past, in part as a consequence of racing to make a new weapon operational as soon as possible, unproven technology had been fielded, only to cause delays, exorbitant expense, and, worst of all, loss of life.

Streamlined management structure would also be tried on the F–15. Deputy Secretary of Defense David Packard had observed excessive layering in management of weapons development work, resulting in delays in making decisions and loss of accountability, especially for failures. Packard was intent on cutting out many of the layers to ensure decisions were made as expeditiously as possible while at the same time delegating far more authority and accountability to the director of the program office.

Program offices are located at product divisions of the Air Force Systems Command, which in turn reports to the Headquarters, United States Air Force. Within the Pentagon, numerous layers of staff are charged with various review and approval responsibilities. At the Air Force headquarters are layers of branches, divisions, directorates, deputy chiefs of staffs, and, of course, the Chief of Staff and his personal staff. In the Air Force Secretariat are, likewise, a host of assistants, directors, deputy assistant secretaries, assistant secretaries, and the Secretary of the Air Force and his immediate entourage. Mirror images multiply above the Air Force in the Office of the Secretary of Defense and out in the field, layer after layer of questions to answer and approvals to reckon with. Secretary Packard wanted major weapons development and production programs protected from exposure to so many potential points of interference.

Packard streamlined management communication channels, keeping just a few key people in the mainstream, specific individuals only, not staffs also, and expressly excluded others who, until the F–15, had routinely engaged in the overseer process. The mainstream elite sensed even greater than usual pressure for the F–15 to succeed under these circumstances. To the extent it existed, that sense of heightened accountability was, of course, one of Packard's principal objectives. However, many excluded from the mainstream would not be disappointed by problems in the program. Problems might substantiate claims that their contributions were essential and irreplaceable.

Inter-Service rivalry was another source of pressure to succeed. Although the notion of another common aircraft had been dropped when OSD approved development of both the F–15 and the Navy F–14, the threat of resurrecting that idea always remained. The Air Force and Navy knew that, if either program encountered difficulty, some observers would quickly recommend that both Services use the one program which seemed to be working. The Air Force always had to check its "six o'clock position" to figure the Navy's whereabouts. Throughout the program, the pressure to succeed or face the prospect of accepting another Navy plane into the Air Force inventory provided a strong incentive.

PROGRAM OFFICE PRINCIPAL

When a new weapon system is to be developed and produced, an office, called the System Program Office or

SPO, is established to shepherd that program through the labyrinth of approvals, tests, negotiations, reviews, and countless other mandatory way points. The director of the SPO becomes the number one person in the Government accountable for the success or failure of that program. The SPO director is accountable for the consequences of all the day-to-day program decisions, budget preparations, contract modifications, schedule adjustments, cost changes, training, logistics preparations, public support, and so forth that *in toto* make the management of weapons systems development such an incredibly complicated and demanding task. Obviously, selection of the director for the F–15 SPO would be one of the most important decisions affecting the outcome of the program. For that job Packard selected General Benjamin N. Bellis, a man uniquely well qualified to accept the pressure of the assignment.

Bellis is the principal author of the 375-series of Air Force regulations on how to manage the development of weapon systems. In the mid-1960s, training films of Bellis (as a lieutenant colonel), lecturing on the process, were shown to program offices and support staff at product divisions to indoctrinate the right way to acquire weapon systems. Moreover, Bellis had been selected in 1965 to direct the highly classified, and highly successful, program to build the SR–71, the most superior aircraft ever built anywhere in the world in terms of speed and altitude. Unfortunately, both it and its interceptor version, the YF–12, were too expensive to build in quantities adequate to counter the anticipated numbers of enemy aircraft. Nevertheless, Bellis' success in his four years as SR–71 SPO director gave him impeccable credentials, and his experiences gave him strong beliefs in how systems management should be conducted properly. Some of the management schemes for the F–15 SPO and the F100 Joint Engine Project Office (JEPO) are attributable to what Bellis concluded from his years with the SR–71 program.

Bellis, too, was an advocate of streamlined management. He had lived under such a structure for security purposes in the SR–71 program. The general accounted to very few superiors and had been free from the typical "staff meddling" with which unclassified programs often feel burdened. Bellis wanted streamlined management as much as did Packard.

Bellis also wanted responsibility for management of the entire program, including the engine. He insisted that the JEPO be part of his SPO. That organizational alignment was not readily acceptable to the Navy, but Bellis believed he had to have complete responsibility for all aspects of the program. If he was to be truly accountable for achieving an initial operational capability by November 1974, he believed he needed authority over every aspect of the program. The engines would be key to the success of the program.[14]

Engine compatibility with the airframe inlet had been a problem with the F–111. Bellis, determined to avoid that problem in the F–15, wanted to manage the JEPO as part of the SPO while McDonnell Douglas was the single contractor legally bound to have Total System Performance Responsibility. Under this contract provision, McDonnell Douglas, the prime contractor for the F–15, had total responsibility for integrating all the F–15 subsystems, such as the engine, into the complete system and assuring the integrated product met all performance requirements.

General Benjamin N. Bellis was Director of the F–15 System Program Office.

Bellis succeeded in gaining the organizational structure he wanted, and the F–15 office became a "Super SPO" on 19 October 1969.[15] The general handpicked much of the cream-of-the-crop in the Air Force for the engineers, buyers, subsystems experts, testers, and the like to man his organization. Two months later the JEPO was formally established in the SPO.[16] The F–15 thus became the first SPO to have total responsibility for its engine.[17]

JOINING WITH THE NAVY

In December 1967, the Air Force and Navy had agreed to work together to develop a high performance afterburning turbofan, known as the Advanced Technology Engine.[18] However, although both Services agreed in principle to join in a cooperative effort, such understanding proved elusive in matters of specifically how the common engine would be finally selected for the F–15 and F–14. Lack of progress in resolving the management issue was noticeable in the Office of the Secretary of Defense. In particular, Dr. John Foster, DDR&E, who had pushed the idea of a common engine through to approval, grew upset, as the Air Force and Navy seemed incapable of agreeing on how they would work together.

The Air Force insisted on total subjugation of Navy involvement to Air Force management authority and acquisition procedures. The Navy feared trouble in such an arrangement because the engines were common but not identical. Though the engine core was to be the same for both, the F100 (Air Force version) would be designed for less thrust and less weight than the F401 (Navy version). The Navy insisted on equal participation. It wanted a joint executive committee with equal Air Force and Navy representation to control the fundamental program decisions.[19] However, the Air Force saw management by committee as unworkable.

In April 1968, Foster designated the Air Force as executive agent for the Initial Engine Development Program, the prototype competition, but this decision only postponed tougher decisions. Who would select the final design and how would both Services initially manage the program and the selected contractor? Neither Service seemed willing to budge significantly from its position, and OSD grew increasingly impatient.

In August 1969, Foster warned the Air Force and Navy assistant secretaries for research and development that unless they resolved their differences and submitted an acceptable management plan to OSD quickly, he would rescind his earlier delegation of source selection authority to them. The Services had no independent authority for acquiring the engine. All acquisition authority for major acquisitions is vested in OSD and normally delegated in writing on a case-by-case basis to each Service. But, since OSD is not required to delegate, Foster's impatience could have led him to take on this major acquisition responsibility himself.

In spite of the worrisome prospect, the Services nevertheless were so entrenched in their positions that disagreement continued. The matter finally reached Packard to be resolved. While contention over this issue had been delaying management plans, work on the prototypes had continued, and decision-time for the final design selection was fast approaching. A final decision on management responsibilities was needed urgently.

Packard directed the Air Force to continue as the lead Service with no joint executive committee. However, based on the unsuccessful results from attempts by the Services to cooperate thus far, Packard warned that he would take back source selection authority if the Services were unwilling to pick a winner. In his memorandum of 10 December 1969 to the Service secretaries, Packard specifically designated the Commanders of Air Force Systems Command and Naval Air Systems Command as the Source Selection Authorities for the F–14/F–15 engine program and made the decision on fundamental management arrangements. The issue had escalated all the way to Packard to settle, and that didn't bode well for future cooperation between the Services.

The Navy struggled with the propulsion restrictions of the F–14. Shown, an F–14 Tomcat takes off with full afterburner from the USS Independence. *(A US Navy photo, photographer unknown.)*

2 ENGINE DEVELOPMENT PROGRAMS

On the previous page, the SR–71, powered by a Pratt and Whitney-developed engine. (An official US Air Force photo, T. Sgt. Jose Lopez, photographer.)

ONE OF THE FIRST TASKS of the Air Force, regarding the technology, was to find out what industry could propose to satisfy the requirement for a relatively lightweight but very powerful engine. The Air Force knew that some key technology had been developed under the Advanced Turbine Engine Gas Generator program run by the Air Force Aero Propulsion Laboratory at Wright-Patterson Air Force Base in Ohio. Advanced components had been brought together to see how effectively they would work as a system. Furthermore, the major competing companies in the industry had developed portions of the engine capabilities needed for the F–15, in response to some earlier R&D work on other projects.

THE EARLY CONTRACTS

The Services wanted to benefit from whatever work had been done in this area of high-thrust engines and to select from all possible sources the most likely proposals to succeed in meeting the Government's requirement in all respects. On 8 April 1968, the USAF sent Requests for Proposals (RFPs) to General Electric, Pratt and Whitney, and the Allison Division of General Motors. The RFPs explained requirements and selection criteria in detail. From these three companies, considered to have potentially the technical and production capabilities to meet the magnitude of the requirements, the Air Force would select two in the Initial Engine Development Program, based on the merits of their proposals, to build two prototype engines (one for each Service). By contract each of the two had 18 months to demonstrate their grasp of the technology.

In August 1968 GE and Pratt were selected. Each had benefited from work on earlier programs to demonstrate engine technology for the Advanced Manned Strategic Aircraft as well as the once planned US/West German V/STOL fighter.[1] Pratt also had accrued major advantages from its design work on the engines for the SR–71. Moreover, Pratt had also benefited from its unsuccessful attempt to win the engine contract for the C–5 and the, by then abandoned, Boeing supersonic transport.[2]

On 17 February 1970, based on hardware evaluations, the Air Force selected Pratt to provide the engines for the F–15, mainly for what was considered its better understanding of the engine/inlet compatibility phenomena so terribly troublesome to the F–111. Pratt seemed to know how to control the mechanism altering fan/compressor pumping characteristics in a distorted inlet flow field. According to a history of the F–15 engine development:

> Basically it was felt General Electric had a better engine structurally and control-wise but a higher risk fan in the stability area. The Pratt and Whitney aircraft engine was assessed to be considerably heavier and lacking a real control system but had a fan/compressor with more potential to meet specifications. Pratt and Whitney Aircraft also demonstrated a greater understanding of engine/inlet compatibility phenomena.[3]

General Bellis would later reflect that Pratt had learned how to compensate for airflow distortions through its experience in developing the SR–71 engine and that this was the key technical difference that influenced the selection of Pratt's design.[4]

So, on 1 March 1970 Pratt won the contract, with future production options, to design, develop, and test the F100 engine for the Air Force and the F401 for the Navy. Thus ended the competition to provide engines for the F–15. Moreover, the contract looked good and tight with tough provisions imposed to keep Pratt in line. Under the demonstration milestone concept, for example, the contractor had to complete satisfactorily a Military Qualification Test (MQT), also often known as the Mission Qualification Test, prior to entering production in mid-1973. Simply put, there was no Government money for production until after a successful 150-hour endurance test. The Air Force wanted proof that technology was really in-hand before making a binding financial commitment for thousands of engines.

Besides the MQT, the Air Force had another reason to feel confident about the contract with Pratt. Even if, in spite of all the testing, deficiencies in the engines later came to light, the Air Force still had protection under the correction of deficiencies clause. Under the terms of this seldom included clause, Pratt was required to correct at its expense any subsequently discovered deficiency stemming from its design, workmanship, or material. In addition, based on expected Air Force and Navy long-term requirements for engines for the F–15 and F–14, the contract included specific prices for a range of quantities in the production options. Thus, while still competing with GE under the IEDP, Pratt had locked itself into prices for future production options even though Pratt was actually the "sole-source" for the engines. The Air Force had seemingly, neatly carried forward the benefits of competition to a time when the technology would have been demonstrated and exact quantities needed for production had been determined.

The F100 program started under pressure to succeed and with the belief that every contractual measure then known had been incorporated to ensure that success. Competition had evidently found the best source. Now the task was to stay on the charted course.

For GE, though, all was not lost. Three months after Pratt captured the contract for the F–15 engines, GE won the contract for the B–1 engines. This sharing of the business for high-thrust jet engines, particularly without a lapse in funding GE's technology team, let the Air Force reap substantial and totally unanticipated benefits years later.

ANOTHER ROUND WITH THE NAVY

The original contract test goals for both engines included a Preliminary Flight Rating Test (PFRT), a 60-hour durability requirement, and the Military Qualification Test. The PFRT was scheduled for February 1972 and MQT for February 1973. However, by the summer of 1971 Pratt slipped the Navy's PFRT to December 1972 and MQT to

June 1973,[5] dissatisfying the Navy with a slower pace of testing components. Although Pratt felt reducing component testing would put overall testing back on schedule without loss in quality, the Navy disagreed. The Navy wanted no extra risks and proposed delaying the MQT three to six months.

STRONG ARGUMENTS ON EACH SIDE

Component testing involves taking certain components out of the engine and running them individually to verify performance. Based on its experience, the Navy was committed to testing each part before testing the assembled whole. Pratt maintained, with some justification, that if the whole works properly, then each of the parts must also. Using such reasoning, Pratt argued to reduce component testing considerably: for example, to cut the fan testing from 900 to 488 hours, the high compressor from 1,290 to 804, and the combustor from 900 to 400.[6] Meanwhile, the Air Force worked eagerly toward a November 1974 initial operational capability (IOC) and believed Pratt could reduce the component testing and still complete MQT in February 1973. The Navy, feeling no urgency to meet the troubled and doubtful F–14B schedule, had no pressure to move forward as quickly since it had a set of problems different from those facing the Air Force.

The Navy, OSD, and Grumman were in a struggle over the F–14. The Navy contract with Grumman was the same type of Total Package Procurement that caused the colossal problems with Lockheed: the C–5 subpar performance and huge cost overruns. Only support from the Federal Government had saved Lockheed from financial disaster; Grumman was rapidly headed in the same direction. The Nixon administration, eager to avoid another embarrassment, would not pour more money into the F–14 program; an alternative was to reduce successively the numbers of aircraft purchased with relatively the same amount of money. So, while the F–15 airframe development kept pushing the F100 engine toward maintaining schedule, the F–14 airframe development had the opposite effect on the F401 development. Because the Navy was willing to slip MQT, it had slack in its schedule; it could go slower.

In testimony before the Senate Armed Services Committee on 28 April 1971, Admiral Connolly, Deputy Chief of Naval Operations, prophetically warned:

> Experience has shown us over and over again that the engine is the longest time thing to get the way it has to be and we have a lot of experience in this and we really think that time will be required for those tests and that we will be better off and safer not sitting there waiting for the engine [for the F–14B].[7]

The Navy began to back away from the F–14B and the F401. Actually, it started scaling down as a program in November 1970 when, because of F–14 funding cuts, it cut engine production from 179 to 69 units for fiscal years 1972–74.[8] This quantity reduction required a new pricing formula in the contract with Pratt. Therefore, a contract modification had to be negotiated with Pratt, and, of course, the prices were higher. Fixed costs, for example, were spread over a small number of units; consequently, unit costs increased. Then in the spring of 1971, the Navy cut the engine order to 58 in line with further slow-downs in the F–14 program.

But delays in component testing were not the Navy's only trouble with the F401. Among the technical problems were first-stage fan blade flutter, fan stator assembly flexing problems, titanium fires, fourth-stage turbine failure, and augmentator rumble.[9] These problems with the technology and the testing schedule discouraged the Navy greatly. When such difficulties were added to the F–14B problems, the prospect appeared very bleak. The Air Force had an inkling of what was coming, particularly after noticing the Navy's budget for the next fiscal year included no money to send to the Air Force for the Navy's share of the program.[10]

On 22 June 1971, Packard chaired a Defense System Acquisition Review Council (DSARC) to evaluate the F–14B status. The result: another delay for the F–14B. In his summarizing memo to the Secretary of the Navy, Packard included guidance on the engine program:

> The F401 engine development program should be structured so that the engine design problems are solved before any production decision on the F–14B. The Navy should expeditiously inform the Air Force of its development and procurement plans for the advanced technology engine to minimize impact on the F–15 program.[11]

One week later, the Commander, Naval Air Systems Command, wrote to the Commander, Aeronautical Systems Division, at Wright-Patterson Air Force Base, and confirmed the Navy would not join the Air Force in ordering production of the engine. The impact of the decision on the Air Force was substantial, potentially crippling. The production option with Pratt could not be exercised by the contractually required date, 1 September 1971. Without Navy participation, the quantities of engines still required fell short of the minimum range of requirements which Pratt had been guaranteed in the competition with GE. Therefore, the production segment of the contract with Pratt could not be executed.

On 15 July 1971 the Joint Engine Project Office asked Pratt to reconsider its engine production program since the previous, competitively determined prices were unrealistic after the Navy's decision not to exercise the production option. Subsequent renegotiation with Pratt, now under sole-source conditions, resulted in a net cost increase to the F–15 program of more than half a billion dollars—$552 million,[12] hardly the minimum impact Packard hoped for.

Why was the increase so steep? Analysis showed, for one thing, that the Navy had been first on the learning curve. This meant the Navy had been scheduled to receive delivery of the first engines, ''technically'' more expensive because of the relatively greater inefficiencies in producing engines by the newly organized work force. In addition, Pratt had just negotiated all, or most, of its subcontracts. Since the negotiated prices were based on larger quantities than would now be required, all subcontracts had to be renegotiated. Furthermore, many of the charges for overhead and tooling, previously shared by the Services, would now have to be borne just by the Air Force. However, the most significant change was that now renegotiation was held in a sole-source, rather than competitive, situation.[13] The Air Force had no apparent alternatives if it still wanted the engines, especially if it wanted them by November 1974.

Instead of breaking the contract and paying so much more for their engines, the Air Force initially considered exercising the original production option by simply buying enough additional engines for Air Force future needs, thus bringing the quantity ordered up to the minimum quantity guaranteed to Pratt. However, the Air Force lacked both the mandatory congressional authorization and the funds to exercise the option. To complicate the predicament further, legal technicalities also may have prevented the Air Force from substituting more F100 engines for the deleted

F401 engines.[14] As a direct result of the myriad of problems brought on by the Navy withdrawal, the F–15 program, which the Air Force had so keenly wanted to be a trouble-free, nearly perfect example of military weapons acquisition at the very best, was saddled with an overnight cost increase of over $500 million. The reasons were explainable, but, over time, observers tend to remember increases more than explanations.

Of course the Air Force's immediate problem was finding money to pay for the increase. All of the management reserve and budget for engineering change orders probably was lost in the reprogramming necessary to fund production of the F100. The SPO would have much less, if any, slack in the budget to accommodate future changes. Naturally enough, when recommendations for improving the engine came forward through the JEPO to the parent SPO, it had to be much more difficult to get budget approval to incorporate changes because of the Navy's withdrawal from the program. How many good ideas, how many chances to solve future problems with the F100, were disapproved because of this sudden, unexpected imposition of severe constraints on program finances?

Eventually the Navy was allowed to continue the F–14 program but with the older, less powerful TF30 engine. Someday the Navy hoped to replace it with the F401. Yet, the cost of such a retrofit program made it out of reach. For many years, the capabilities of the F–14 would be restricted by inadequate propulsion.

QUALIFICATION PROBLEMS

Edward C. Westwood describes the F100 program as "a constant struggle of man against nature."[15] Problems with the test engine confirmed the enormity and wide variations of the technical puzzles to be solved. The fundamental problem, continually confronted, was how to make the F100 rugged enough to generate 3,000 pounds more thrust than its predecessor while at the same time cutting the weight of the engine by 1,000 pounds.[16]

Fan blades are an example of this basic challenge. How could they be made to withstand the stresses generating the boost in thrust? Blades often fluttered, then broke, and sometimes caused fire. Reinforced shields, ablative blankets, and fire dams could contain damage. Containing damage wasn't the goal, though. What materials and design changes could prevent blade failure without adding weight? Combination after combination was tried.

Another serious fan problem was rotor-to-stator rubbing. Titanium fires sometimes erupted from the friction. After a while, a nickel alloy substituted at certain vulnerable points eliminated the titanium-on-titanium rub. However, another serious technical obstacle still had to be passed.

Time proved the Navy accurate in predicting delays completing the qualification tests. Perhaps the Navy had also been right in arguing for more time to complete component testing leading up to the system test. In any case, 1973 was not a good year in the history of F100 development.

In April 1972, Congressman Flynt, at a hearing of the House Appropriations Committee, asked General Glasser, Air Force Deputy Chief of Staff for Research and Development, "Do you consider the Air Force F100 program mature enough for MQT in February 1973 when, only two months prior to this date, the Navy, with essentially the same hardware design, will only be ready for PFRT?" In his response, General Glasser stated, "The maturity of the F100 engine at MQT will be greater than that experienced in other weapon systems programs at an equivalent stage.''[17] That was the positive attitude about the program exuded just about everywhere in the Air Force. However, the task ahead was very difficult.

The qualification test would be conducted not on an aircraft but on a test stand, in a test chamber rigged to simulate, for example, altitude conditions. The 150-hour test would be run at sea level, at specified altitudes, and at varying Mach numbers.

The first F100 to attempt the official MQT failed on 26 February 1973 after about 60 percent of the test had been completed, because of problems with the fan and turbine blades.[18] Then later, during an altitude test with another F100 engine at Tullahoma, Tennessee, on 14 March 1973, an explosion caused extensive damage to the engine and to the test chamber. Analysis of that engine failure arrived at the SPO shortly before the second attempt at the MQT was scheduled to start. The data from Tullahoma pointed to a problem with the stator. Under stress, it moved forward, touched the rotor, and ignited a titanium fire.[19] These test failures weighed heavily on General Bellis as he considered the importance of the upcoming second MQT attempt.

Aside from problems with the F100 engine, the F–15 program was moving steadily towards its November 1974 completion date. In fact, the aircraft production schedule was progressing very well, without any delays. Only the engine raised concern. Because the demonstrated milestone concept required successful completion of MQT before funds could be released to start engine production, the problems in completing MQT were threatening to delay engine availability which, in turn, would delay F–15 deliveries and operational capability. Not only would engine delays affect the IOC, they would be expensive. In the contract with McDonnell Douglas, the Air Force was obligated to ensure the engines were provided at the specified times for installation in the aircraft. Costs incurred by McDonnell Douglas as a consequence of delays in getting the engines would be chargeable to the Air Force. Therefore, MQT problems also threatened the cost and schedule objectives toward which the SPO had been so aggressively striving for three years. It was a crucial moment.

For both professional and personal reasons General Bellis was especially mindful of his original tasking directly from Packard at the inception of the program. Packard had charged him with responsibility for success or failure of the program and had vested in him, particularly through streamlined management, broad program authority. During his conversation with Packard, Bellis had mentioned he wanted tougher requirements for the endurance run than standard military specification called for. Packard had no objection but cautioned Bellis to be flexible during the program. Packard especially instructed Bellis not to do anything careless such as the decision, made by others earlier, to hang on to the C–5 flotation landing-gear requirement even after dropping the requirement for landing the aircraft on dirt and grass. The message had seemed clear to Bellis: Don't hold on foolishly to self-imposed requirements.[20]

The SPO felt the conditions causing the stator flexing problem were encountered under high-altitude, high-Mach-number conditions. However, Pratt would need too much time to confirm the analysis and then design, fabricate, and install the correction. Since the MQT requirement was stricter than standard, why not run the test without burdensome conditions? Bellis agreed and waived the requirements. Further, he notified Pratt that completion of the modified test would satisfy the contract requirement for engine qualification prior to production.

Bellis made his decision on 30 March 1973, but he delayed telling the Secretary of the Air Force and the Chief of Staff of the decision until 14 April and OSD until 16 April. The second test, as modified, was successfully completed on 17 April. Despite the success a furor erupted. Bellis's superiors were livid that they had neither participated in a major program decision nor, absent that participation, been informed of the decision promptly. Several were embarrassed that they had misinformed Congress in recent hearings, even though innocently, about the program status.

An angry Congress held hearings in the Senate and the House to review every particular of the decision. Fingers pointed to Bellis as the culprit. Although only General Glasser spoke to Congress in Bellis's defense, Bellis still feels that he had the full support of his Air Force superiors throughout the review.[21]

At first glance the change in test requirements didn't seem that significant. Instead of testing the engine at Mach 2.3 and 37,000 feet, the test was run at Mach 2.2 around 40,000 feet. However, inquiry revealed these seemingly minor deviations actually made a very substantial difference on the total pressure at the inlet face of the engine:[22]

	Pounds per Square Inch	*Inches of Mercury*
Mach 2.3 at 37,500 feet (the contract requirement)	34.30	69.8
Mach 2.2 at 37,500 feet	29.66	60.39
Mach 2.2 at 40,000 feet (the modified requirement)	23.89	48.4

The pressure had been reduced by about a third at a critical point in the final test of engine worthiness before production. Consensus condemned the waiver. The publicity was bad.

Moreover, change in OSD personnel at the top levels aggravated Bellis's situation. Packard, with Foster as DDR&E, had approved and enforced streamlined management with greater responsibility and authority to program managers. However, those two officials were gone by 6 April 1973, replaced by William Clements and Malcolm Currie, respectively. Although in theory the same management policies continued under the new leadership, the bond Bellis had with Packard from hours of discussions in quieter times was missing.

Clements sent letters of explanation to the key congressional committee chairmen. So did Secretary of the Air Force Seamans. Since Bellis had obligated the Government to accept the modified test as meeting the contract requirement for engine qualification, an accommodation had to be reached with Pratt without reopening the whole contract to renegotiation and another hefty increase in price. Though Pratt received $38 million, a partial allotment, to start manufacturing engines, the 150-hour test would have to be run on the improved engine under the originally required conditions. Until the test had been completed successfully, no additional funds would be allotted, and any engines accepted to fill holes in F–15s would be called "pre-production units."[23]

Through the crises of early 1973, the Air Force managed to keep Pratt and McDonnell Douglas working toward November 1974 without reopening either contract. However, the issue over the MQT waiver would not go away because the engine kept having problems trying to pass the test. In July 1973, the test was fouled by contaminating dust from the test cell. The test was restarted but failed again, after 132 hours, on 22 August following a turbine rotor failure. After modifications, the test resumed on 13 September only to stop again two days later to replace three broken turbine blades. The blades had overheated badly and fractured under great stress.[24] Finally, after false starts, the qualification test was completed in October 1973.

Full qualification opened a relatively clear path to full production, on-time deliveries for installation in F–15s, and initial operational service at Luke Air Force Base, November 1974. General Bellis left the SPO in 1974 and in 1975 was commander of the 17th Air Force in Europe, preparing to introduce the F–15 into overseas operational inventory. Without doubt, though, the waiver incident tarnished the otherwise sterling record and reputation for technical management of the F–15 Systems Program Office.

BUSINESS PROBLEMS

By 1975 the Government had lost many of the benefits it had achieved from the head-to-head competition in the IEDP. The price advantages of competition disappeared when the Navy dropped out of the program. Lower quantities meant higher unit prices as fixed costs were spread across a smaller production base. Furthermore, the experience with the qualification test brought out weaknesses in the demonstration milestone concept. The old days of continuing development throughout the production cycle returned.

The program schedules, with development and testing continuing as the first engines were being manufactured, added significant risk of future technical problems and high costs. Even the correction of deficiencies clause was dropped before the year was out.

Financially, Pratt was in trouble in the contract. Its corporate parent, United Technologies Corporation (UTC), had recently been stung in the commercial engine business. Pratt was suffering, in part, from the abnormal escalation of prices in the economy. During 1975 Pratt filed several hundred million dollars in claims against the Government for economic price adjustment to account for inflationary pressures beyond Pratt's control. Pratt's target costs were already at or above its guaranteed not-to-exceed prices for production options.[25] Understandably UTC and Pratt were anxious over the correction of deficiencies clause. From that clause alone the financial guarantees to the Air Force could cost Pratt dearly. So when the price adjustment claims were under negotiation, Pratt shrewdly added the correction of deficiencies clause to the agenda. In the give-and-take of resolving individual issues, Pratt hoped to delete this onerous clause.

Also in 1975, the Air Force selected the General Dynamics F–16 design in the lightweight fighter competition. And, since the Air Force planned to power the F–16 with Pratt's F100 (one per aircraft, rather than two, as in the F–15), it found itself totally dependent on Pratt for all high performance fighter engine needs as far into the future as anyone could see.

Although the Air Force would soon focus increasing concern on the extra vulnerability to catastrophic failure inherent in a single-engine, rather than twin-engine, fighter, the more pressing concern at the time was Pratt's financial health. Pratt claimed economic hardships were practically driving them out of business. Such a calamity befalling the single source for advanced fighter propulsion made strong impressions on Air Force leadership. Pratt must be saved. Expediency drove the Air Force to seek a new agreement with Pratt just as surely as expediency drove Pratt to bring the correction of deficiencies clause to the negotiating table.

Perhaps unexpectedly Pratt found some allies in the military. The safety net initially expected from the clause was no longer credible to those close to the contract in the Air Force. The uncertain status of the specification, the large number of engines already delivered, and the blurred view of accountability in the controversial qualification procedures were among the factors considered. Believing it really wasn't giving up anything and feeling it was responsive to Pratt's needs, the Air Force agreed in September 1975 to drop the correction of deficiencies clause as part of an overall, out-of-court settlement of the Pratt claims. In the settlement, Pratt agreed to fix a number of specified, existing deficiencies and to accept addition of a standard supply warranty provision. Pratt, of course, was pleased with the arrangement. A Pratt executive privately told General Bellis the company would never again sign up to a correction of deficiencies clause.[26]

Without the correction of deficiencies clause, Pratt profited enormously. Pratt had both the F–15 and F–16 engine business wrapped into one contract with noncompetitively determined prices. Moreover, the Component Improvement Program (CIP) was in full swing. Swelling corporate coffers could well support continuing corporate expansion.

THE COMPONENT IMPROVEMENT PROGRAM

The Component Improvement Program is a Government-funded, somewhat open-ended part of a contract. It exists because no matter how thoroughly developers attempt to find and solve engineering problems during preproduction tests, some problems are discovered only through operational service. Sometimes years pass before conditions, neither anticipated nor contractually covered in the original scope of work, create new problems requiring prompt solutions. The CIP is sustaining engineering support intended to cover improvements to such performance characteristics as reliability, maintainability, durability, and operability after the equipment has entered full-scale manufacturing and even operational use. The program begins with the contractor submitting a list of proposed, worthwhile efforts, often made in response to problems and concerns already identified by the Air Force. Normally, the list is larger than can be afforded so the Air Force evaluates, selects, and rank orders the most important, affordable tasks. Of course, flight safety concerns or conditions which might ground aircraft receive the highest priority.[27] Some CIP work stems from the way the development phase is conducted. The more concurrency and compression of the ideal development schedule, the more likely that incomplete analyses and tests will yield problems later on. Obviously, these were the circumstances in the F100 development.

The engine contractual work started after the airframe development had already begun. It was always being prodded to catch up. Years later, Dr. William Perry, Under Secretary of Defense for Research and Engineering, would state in a congressional hearing, "In retrospect, it probably would have been wiser to hold the engine back from operation until testing was further along and until it had achieved a degree of maturity."[28] That might have been the technically prudent judgment, but that decision almost certainly would have caused delay in meeting the initial operational capability. The consistent decision was always to press forward. Even though F100/F401 development costs rose from $117.45 million in 1970 to $1,003.9 million by 1975,[29] the F100 entered production before it had been completely developed and tested. Therefore, it could have been expected that CIP tasks would be used to complete the development phase as well as to perform the usual production engineering.

The cost of the CIP to the Air Force can be great. The 1981 fiscal year President's Budget identified a cost for the F100 CIP of $681.2 million from 1974 through 1985. By the end of 1980, $419.3 million had already been obligated by the Air Force.[30] As operational use of the engine accelerated and problems occurred with increasing frequency, the Air Force would often wonder why more wasn't being accomplished with all this money.

3 PROBLEMS WITH THE ENGINE

On the previous page, the F–15, shown with nozzle flaps removed from the F100 engines. (An official Air Force photo, Theodore J. Koniares, photographer.)

THE F–15 WAS EXTREMELY WELL RECEIVED by the pilots. They were delighted with the maneuverability, acceleration, and rate of climb of the aircraft, performance characteristics largely made possible by the two F100s providing the propulsion. With this far superior performance came development of new air combat tactics to capitalize on the new capabilities, and with the tactics came additional flying hours for training. However, as flying hours accumulated rapidly and pilots practiced darting, evasive movements again and again, major problems arose. One of the first was occurrence of a high rate of stall stagnations.

STALL STAGNATIONS

An engine stall is a momentary hesitation in engine operation caused by a disturbance in the air flow and resulting in an aerodynamic stall of the compressor blades. If the stall is not self-recovering through immediate adjustment of the air flow, it is called a stagnation. The pilot must then shut down the engine and restart it.[1] By July 1977, 223 F100 stall stagnations had occurred in the F–15—76 percent of them while the afterburner was in use.[2] As the occurrence of stall stagnations kept increasing, jubilation over the F–15's performance waned, and the reputation of F100 became tainted.

The Air Force followed two parallel paths to correct the problem. One path sought a permanent solution through design changes. However, until the optimum engineering changes could be found, the parallel approach restricted operation of the aircraft to lessen the chance of encountering a stagnation.

Some of the restrictions dealt with the way the pilot flew the aircraft. Analysis of stall stagnation accidents showed they tended to occur most frequently in the upper left hand corner of the flight envelope. In that situation, operating at high altitude and low speed, rapid throttle movements were more likely to cause pressure spikes that produced the problem in the engine. To avoid spikes, pilots were warned to use "TLC (tender loving care) to decrease the chance of stagnation, particularly when in that part of the envelope."[3]

However, relying on the self-control of the pilots wasn't the only restriction. Mechanics adjusted or detuned engines to lower thrust performance.[4] Called "trimming," the procedure was not as easy to do as the name suggests. The F100 brought with it an unanticipated maintenance burden of troubleshooting and trimming. As the workload on the Air Force maintenance technicians rapidly increased above what had been expected, naturally the engine's reputation suffered further.

But the trimming procedure was only for peacetime flying. In wartime, detrimmed engines would be restored to their originally intended performance potential. The hope was that permanent solutions would arrive before any conflict. However, even if they did not, all engines would eventually be detrimmed and any ensuing stagnations accepted as an additional hazard in war.

As aggravating and worrisome as the engine problems in the F–15 were, concern over the engine's performance leaped upward when the Air Force thought of stagnation problems in the F–16. Even before the F–16 became operational, everyone quickly understood that the risk of stagnation in this single-engine aircraft had far graver consequences. While the F–15 pilot could rely on the other engine if one stagnated, the situation would be far trickier in the F–16. An F–16 pilot would have only three options: quick restart, emergency landing, or ejection.

Already pilots were beginning to fly in ways unsuitable for combat but, nevertheless, essential for prudent, safety-conscious, peacetime operation. Inadequate technology was forcing pilots to fly the engine rather than the aircraft. Instead of thinking of the F–15 as a whole and flying that system to its limits, they had to pamper the engine. In evasive, tactical maneuvers instead of keeping their heads up, focused on targets and threats, pilots more frequently looked down, distracted by the need to check engine gauges inside the cockpit.

While these restrictions limited aircraft operations, Pratt worked on permanent solutions; however that fix-it effort was not going smoothly. Problems between the Air Force and Pratt on this issue previewed the even more intense disagreements to follow.

An F–15 Eagle undergoes an engine change (above); an airman performs F100 (F–15 engine) maintenance (below). (Official US Air Force photos; photographers unknown.)

One of the most disturbing aspects concerned the Air Force perception that Pratt was more interested in generating profits through contract changes than in making the engine perform properly. Pratt lawyers contended Pratt had no contractual obligation to eliminate stall stagnations without receiving more money since the F100 had been designed and qualified to the Air Force specifications.[5] After all, the Air Force had accepted the design for production. If the Air Force later wanted to add requirements to change the performance, that change was an additional tasking, not a part of the original contract, and subject therefore to additional charge.

The Air Force leadership felt that if they went to Pratt to discuss technical problems, instead of meeting engineers and other F100 program people, they would meet Pratt attorneys.[6] At one point, in order to make some headway in solving the engine performance issues, the Air Force met with some of Pratt's best engineers at a neutral site, unbeknownst to Pratt's top management and lawyers. The Air Force knew that the good solutions of Pratt's engineering staff were not getting past their management as quickly as the Air Force wanted because of haggling over business terms. So at an Atlanta airport hotel, the Air Force and Pratt met quietly and produced some much needed quality, independent research.[7]

Of course, the design solutions were difficult to incorporate in the midst of the ongoing high rate of F100 production. In a less time-pressured development period, perhaps someone might have spotted the stagnation problem before engines were manufactured in volume and installed in operational aircraft. Unfortunately, since that did not happen, the problem was much more complicated and costly to correct.[8] Manufacturing operations halted to incorporate changes, and good parts had to catch up to bad parts in the field.

Eventually, however, three fixes were made, two alone for the afterburner. Upon sensing a stall, engine monitoring devices reduced afterburner fuel flow to a minimum and, at about the same time, opened the afterburner nozzle to decrease the afterburner pressure pulses.[9] The third fix applied only to the F–16—the proximate splitter. This mechanical feature divides the fan air flow between the bypass duct and the high-pressure compressor, reducing the likelihood of a pressure disturbance from the fan duct flowing back into the engine core.[10] Though considered necessary as extra protection from stagnation in the F–16, the splitter wasn't essential in the twin engine F–15 and, therefore, was not to be used in that aircraft.

Rapid results came from these corrections. Although in early 1976 the stall rate was averaging 9 per 1,000 engine flying hours, by the end of 1977, the rate was down to 4.1 stalls per 1,000 flying hours.[11] Because the problem had been discovered quickly enough in F–15 operations and corrections made, despite some perceived footdragging by Pratt, the F–16 experienced relatively few stagnations. In fact, when General Allen, Air Force Chief of Staff, briefed on the engine to the House Appropriation Committee in February 1979, he reported that stall stagnation had been "nonexistent in the elements of the flight envelope that are of concern."[12] One year later, in a hearing before the same committee, the Secretary of the Air Force added that the low number of stall stagnations in the F–16 was attributable to a different engine inlet and engine installation.[13]

Nevertheless, stall stagnations had not yet been eliminated and the problem attracted substantial, high-level visibility in the Air Force and the Office of the Secretary of Defense. The experience with stagnation episodes, the extra maintenance burdens, and the restrictions on operations gave rise to the first Air Force thoughts that an alternative to the F100 might be needed.

In November 1977, General Dixon, Commander of the Air Force Tactical Air Command, sent a message to General Slay, Commander of the Air Force Systems Command, expressing displeasure at having all his tactical capability dependent on one engine.[14] Though GE may have had a hand in instigating the message, Dixon was clearly troubled by recent events. Coming at a time when he was stressing greater realism in combat training, the limitations on aircraft performance worked directly counter to his goals for better preparing his combat pilots.

Dixon's message prompted concerted inquiry for a solution in Air Force Systems Command. General Robert Mathis, the command's vice commander, who had also been a director of the F–15 SPO after General Bellis, took particular interest in the problem and its solution.[15]

As more time passed other high officials had first-hand experience with the F100's difficulties. Dr. William Perry, Director of Defense Research and Engineering in the Office of the Secretary of Defense, went to the Air Force Flight Test Center at Edwards Air Force Base in California in early 1979 for an F–16 flight. En route he read background materials prepared by his staff. Aware that the F–15 and F–16 used the same engine, Perry was disturbed to read that the F–15 had by then made 50 single-engine landings due to F100 stagnations. When Perry checked further with the F–16 pilot at Edwards who was to take him on his orientation flight, he was startled to hear his pilot report on 10 stagnations he had experienced in the single-engine F–16. The pilot tried to reassure Perry by explaining that these incidents were all outside the normal flight envelope. Perry was obviously concerned, but he didn't back out. He took the F–16 ride. His personal experience defined the problem better than statistics on paper. A few months later, during a hearing before the Senate Armed Service Committee, Perry described the F100 failure rate as "unacceptable."[16]

General Slay had a more harrowing experience. While Commander of Air Force Systems Command, he personally flew the single-seat F–15. Just a few days before a congressional hearing on the engine in 1979, and already by then perhaps Pratt's greatest adversary in the Air Force, General Slay experienced a stall stagnation while taking off over Dallas with a full load of fuel and poor visibility. He had to solve this in-flight emergency quickly. Fortunately he did.[17] Thereafter the general spoke on the subject with the great authority of personal experience.

From a multitude of angles, the stall stagnation problem upset the Department of Defense. For everyone from pilots and maintenance technicians to senior people in the department, the shining prospects of the F100 and Pratt faded. The initial glow from the success in fielding such a high-performance system on time faded quickly as concerns for operating and maintaining a dependable system became the primary focus of attention.

THERMAL CYCLES

Historically, up through the qualification of the F100, engine life was measured by its operating time. In tests, the emphasis was on accumulating many hours of engine-on time, particularly at high-Mach numbers, to check the design and durability of the engine. However, even while the F100 was still being developed and manufactured, elsewhere in the Air Force and the propulsion community, engineers were discovering a more important criterion for measuring engine life. As explained by Assistant Secretary of the Air Force Thomas Cooper to the House Armed Services Committee:

> The original 150-hour qualification test required extensive test time at high Mach number (greater than Mach 2) intended to proof test against stress rupture failure. We came to understand through the 1970s that other factors were equally severe in engine usage, namely low-cycle fatigue, or repeated exposure to thermal and mechanical stress as the engine is cycled between cut-off and various power settings seen in this high thrust to weight fighter mission.[18]

A throttle movement from the idle position to maximum power and then to an idle or an intermediate-range position defines an engine cycle. Engineers found that, even more important than the accumulation of operating hours, the number of cycles accumulated was of paramount importance in limiting engine life. As the power settings of the engine change, the temperature and mechanical stresses change also. An engine is relatively cool at idle position. As the power setting increases, engine temperature rises. Parts heat up until, at maximum power, the temperature is at its highest. When the throttle eases back, power reduces, the temperature drops, and the parts cool off again. The more the parts are stressed by the heating and cooling cycle, the more likely it is the stresses will cause distortions to the systems and fractures or disintegration of parts. The situation is similar to the operation of a light bulb, less likely to burn out after being on steadily, even for many hours, than at the moment it's switched on or off. The most likely reason for the filament to break when the switch is moved is the same for engine parts. Whenever the temperature changes, the metal expands or contracts. Over time these contractions have a cumulative effect sufficient to degrade the part.[19]

Of course the Air Force had known for a long time that engine life was influenced by power setting.[20]

However, the full impact of stresses on engine durability, especially on a race horse like the F100, was unanticipated. During the F100 development most technicians felt that the length of time spent at the highest temperature was the most important determinant of stress on engine parts. Thermal cycling wasn't widely understood by the USAF or the engine contractors. Also unexpected was how frequently pilots would push the engine to full throttle position and then pull back to a moderate setting.

The performance of the F100, when it worked, was so vastly superior to any predecessor that it changed combat tactics and the training regimen. Cycles accumulated at far more rapid rates than planned as pilots pushed the aircraft through rapid, evasive maneuvers. Relatively little time was spent at full power or very high Mach number. Much more often, speed stayed below Mach 1.5 with power settings changed often and abruptly as pilots trained to outmaneuver the foe. The figures below show the planned versus actual results of four key usage parameters of the engine as of November 1979.

	Design Requirement	*Actual Operational Use*
Total operational time	3,000 hours	2,000 hours
Full-power	525 hours	235 hours
Mach 1.6 or above	69 hours	10 hours
Full-throttle transients	1,765 cycles	10,360 cycles*

Source: US Congress, Senate, Committee on Armed Forces, *F–15 and F–16 Engine Problems, Hearings,* 96th Cong., 1st sess., 27 November 1979, p. 71.

*Projected in a 2,000-hour operational life.

The almost six-fold increase in full-throttle transients, or thermal cycles, stands out as the greatest single difference between the F100 design requirement and actual experience in operational use. Neither the Air Force nor Pratt anticipated this change in tactics and demands on the engine. Perhaps the change could not have been anticipated. Perhaps no one could have fully appreciated the capabilities and new opportunities provided by the F–15 until fighter pilots could actually get behind the stick of production, not experimental, models and put them through their paces. In any case, as engine breakdowns occurred with increasing frequency, Pratt pointed out the growing gap between design goals and operational usage. Since Pratt had demonstrated through qualification tests that the F100 met the contractually required design specifications, changes to the design, now that the effect of thermal cycles was becoming better understood, would require additional Government funding to increase the scope of the contract.

Although it's very doubtful that the increased volume and rate of accumulating thermal cycles could have been anticipated, the general impact of thermal cycles on engine life should have been known at the time the F100 was being developed. Perhaps it was. An Air Force engineer at Wright-Patterson Air Force Base recalls that when he was assigned to the F100 development program, he quickly noticed the absence of a requirement for achieving any specified structural life objectives or of criteria for measuring fatigue. Having had recent experience with airframe development techniques, he knew that the duty cycle (how the aircraft would be used on a typical mission) was a key factor in determining structural life. For an engine, duty cycle is the same as a thermal cycle. Consequently, the engineer believed it was important to test an engine as an airframe is tested, simulating actual usage. Moreover, problems with other engines, such as the TF30 and TF41 in operational use at the time of F100 development, had detectable links to stress failures of component parts. Accordingly, the engineer urged the engine be required to achieve duty-cycle standards the same as those imposed on the airframe. Unfortunately for the Air Force, this engineer never succeeded in convincing management to change requirements and now believes some of his superiors may have felt that imposing a duty-cycle requirement would have required increasing engine weight to increase durability. Adding weight would have lowered the thrust-to-weight ratio of the engine and might have disrupted the schedule to meet the IOC.[21]

Very few people seem to recall the discussion of duty cycles as vividly as this single engineer. Nevertheless, even if the details of his recollection are imprecise, clearly somewhere in the Air Force the importance of requiring engines to meet specific durability standards for cycles was being recognized at about the same time the F100 was being developed. Pratt was awarded the contract for the F100 in February 1970. Just four months later, in June 1970, GE was awarded the contract to develop the engine for the B–1 bomber. The GE contract included a durability requirement prompted by observation of thermal cycle damage to engines.[22] Yet, the Aeronautical Systems Division at Wright-Patterson Air Force Base awarded both the Pratt and the GE contracts. In the comprehensive coordination of the specifications and contracts for those two major development efforts at about the same time and at the same location, did no one see the difference in the durability requirements and wonder why they shouldn't be the same? Possibly the difference existed because the four-engine bomber could afford the extra weight, resulting from heavier parts to meet tougher durability requirements, while the weight-sensitive fighter could not. In hindsight, thermal cycles should have also been inserted in the F100 design and testing requirement. In January 1976, the Air Force Scientific Advisory Board, in a special report on turbine engines, concluded:

> The proposed usage of the engine should be carefully defined at the beginning of engine development. This definition should include details of the frequency amplitude of throttle movements. AFSC [Air Force Systems Command] should carefully specify a test program that actually reflects the proposed or current usage.[23]

TURBINE FAILURES

As fiery hot gases exit the engine combustor, stationary vanes and turbine blades bear the brunt of the thermal stresses as they direct, and are propelled by, the gases. When the high rate of turbine failures alarmed the Air Force, engineers traced the causes primarily to the stresses on the vanes and blades.

Turbine failure is often a euphemism for engine blowup.[24] When turbine components break, the resulting debris not only destroys other engine components but also is thrown out of the engine where it does even more damage—some of it potentially fatal. Damage done by broken blades is particularly great because the extreme centrifugal forces generated by their rotation in the turbine can propel them out of the engine and through the fuselage and possibly the crew members. Unlike stall stagnations, which a pilot can usually correct in-flight by shutting down and restarting the engine, the damage from a turbine failure destroys the engine. If the damage is only to one engine and the aircraft has two, the pilot can safely return. However, just as with stall stagnation, the greater concern with turbine failures was the likely consequence of single-engine failure. Not only is there no fall-back on which to rely, the F–16 fuel tank is located just above the engine.[25]

Seventy-five percent of the first 54 turbine failures were attributable to overtemperature stresses, of which 13 stemmed from the overtemperatures created by stagnations. Air flow during stagnation was insufficient to cool the engine but was still enough for combustion of the continuing fuel flow. In consequence, temperatures soared. An uneven combustion pattern caused another 15 of the failures. While the average combustor temperature was 2,250°F, certain hot spots spiked up to 3,250°F. This disproportional pattern heated some parts more than others. Over time the imbalance caused parts exposed to this imbalance to break.[26] The problem was particularly acute for blades. Subjected to the thermal overstresses, the blades cracked. Then centrifugal force expanded the cracks until blades broke, creating highly kinetic, lethal projectiles.

Several corrective steps were undertaken. Some were temporary fixes, and were made while long-term solutions, such as new material technologies, were being developed. The engine trimming procedure, which helped to counter the tendency toward stagnations, helped to reduce overtemperatures as well. Durability increased as operating temperature decreased, although the generation of less thrust lowered performance.

Uneven combustor temperature profiles were improved by adding filters to the fuel nozzles to keep debris from jamming the nozzles open. Later, a new spin-type nozzle more evenly dispensed fuel throughout the combustor and created a much more balanced temperature pattern. Thermocouples, which record engine temperatures, were also checked more frequently, and an aural tone, warning of overtemperature, was added to alert the pilot to a dangerous condition.[27] In addition, extra metal, called belly bands, placed around the outside of the engine helped contain wayward fragments. Unfortunately, these containment improvements added almost 48 unwanted pounds to each aircraft, 35.8 pounds for the belly bands plus another 12 pounds for a thicker turbine case.[28]

The flexible borescope greatly aided the Air Force in detecting cracks and other material defects in turbines. This $23,000 fiber optic device allowed a maintenance technician to inspect each internal, critical part of the engine while still installed in the aircraft. The procedure to check the inner workings of the engine, including all 268 turbine blades, dropped from six hours to just two hours with the fiber optics. The findings were impressive.[29] When the Air Force issued a Time Compliance Technical Order for an immediate safety check of the engines with the borescope, 23 percent of the 557 engines inspected were rejected.[30] Once borescope inspections began, incidents of turbine failure plummeted. Between March 1979, when borescoping started, and late November 1979, the Air Force discovered and replaced distressed turbine blades and vanes in 174 engine cores.[31]

The maintenance burden, however, was enormous. The inspection procedure was required on each F–15 engine after it had operated for 100 hours and on each F–16 engine after it had operated 50 hours.[32] Yet, because the results of the borescope inspection were so reliable and because of successful engineering changes stemming from the Component Improvement Program, engine tear-down inspections weren't needed so frequently. Instead of removing, inspecting, and refurbishing an engine after every 900 cycles, that key maintenance point could be extended until 1,350 cycles had been accumulated.[33]

OTHER PROBLEMS

From December 1976 through 1977 another problem surfaced—40 main fuel pump malfunctions resulted in 26

F–15 single-engine landings.[34] These safety problems led to a switch from a vane to a gear-type pump. In addition, in 1979 problems with the afterburner alone accounted for almost 50 percent of the manhours spent on unscheduled maintenance. About two-thirds of the time this maintenance was on the external titanium nozzle flaps which split and cracked under stress, another fatigue problem. The original design intent for the flaps was to reduce aerodynamic drag over the engine; however, when one pilot tried flying without the nozzle flaps, he noticed little change in performance. Thereafter, all F–15s at Luke and Nellis Air Force Bases flew without the flaps.[35] This was probably the most easily solved of all the engine problems, and it did not affect the F–16 because its cleaner aerodynamic lines in the back produced less buffeting of the flaps.

Not so easy was the problem of control system reliability. Complicated mechanisms control the operation of the F100. Over 60 different components must perform precisely to control simultaneously all aspects of the engine. The two components needing most attention are usually the unified fuel control and the electronic engine control. Other control components also have problems, and by the late 1970s, control system reliability defects accounted for over 27 percent of all unscheduled maintenance on the engine.[36] And these problems lingered. Design improvements, such as the Digital Electronic Engine Control, would not be incorporated readily and would become a big feature of much later proposals from Pratt to the Air Force.

INTANGIBLES

The many problems with the F100 cannot be blamed solely, or even predominantly, on Pratt. Nor can they be identified with any degree of precision. Remember that the

F100, rushed into production, was designed and built to Air Force standards, although some were later found less important than other standards (e.g., thermal duty cycles) unmentioned in the specifications. Further, the Air Force flew the engine far more often and in more unorthodox ways than envisioned when it was designed. Furthermore, the Air Force logistical provisioning of spare parts had not kept pace with projections of parts usage. The F100 never had a chance to be operated moderately at first and then gradually eased into full production and operational use. It never had a chance to mature. Everyone associated with it was constantly trying to catch up to the frantic pace of solving problems concurrent with aggressive operational use, which often disclosed even more problems.

In the midst of this frenzy, the stresses on the engineering and business aspects of the Air Force's relationship with Pratt were considerable. On technical matters, the dialogue often seemed to be generally constructive to the Air Force, limited more by the imposition of constraints by Pratt's lawyers and senior managers than by the complexity of the performance problems. On the business side, Pratt's perceived interests in positive cash flow and profits over solving the problems continued to irritate Air Force echelons.

4
THE BATTLE LINES

On the previous page, a high-angle front view of the two-engine F–15. Powered by the F100, this aircraft was one of the first assigned to the European theater of operations. (An official US Air Force photo, Georg Wegemann, photographer.)

PRATT AND WHITNEY

Pratt and Whitney, founded in 1924 to design and build radial, air-cooled aircraft engines, started on a very modest scale, operating out of a leased machine-tool factory. The company's first sale was for just six experimental engines to the US Bureau of Aeronautics for $90,000.[1] Not long after its founding, the company was involved with its first big merger in a corporate history frequently associated with mergers and spin-offs.

After Boeing reorganized in 1928 as the United Aircraft and Transport Corporation, it quickly merged with Pratt and Whitney and other aerospace firms to form a conglomerate in competition with General Motors and Aviation Corporation (AVCO). In spite of this competition, however, during the 1930s congressional concern over the "dovetailed nature of transport service and manufacturing concerns"[2] led to restrictive legislation, antitrust action, and eventual divestiture. The conglomerate was divided into Boeing, United Airlines, and United Aircraft, the latter including Pratt and Whitney, Hamilton Standard, Chance-Vought and Sikorsky. Throughout this period Pratt and Whitney remained in East Hartford, Connecticut, where it started, primarily because of the pool of available skilled labor.

As years went by, Pratt and Whitney generally prospered. Operations expanded, thousands of engines were made, and management was savvy enough to ensure Pratt was one of the relatively few engine manufacturers to survive the instantaneous phasedown of production after World War II. At that time, one of the company's wisest decisions was to focus its attention on the emerging market to supply the commercial airlines with jet engines. General Electric actually had a technological advantage in getting into this market, but Pratt concentrated its resources and focused its marketing skills more effectively on the new civilian market. Consequently, when Boeing and Douglas, which were building large airframes for the airlines, evaluated the various competing engine manufacturers, they found Pratt most responsive in most cases and selected Pratt engines to power their aircraft. By the end of the sixties, Pratt had captured over 90 percent of the civilian market.[3]

Later, when Pratt succeeded in winning the competition for the F–14, F–15, and F–16 engines, the company seemed to have gained an unbreakable control of the military market as well. The F–16 contract was the sweetest of all. The Air Force planned to buy 1,383 aircraft for $13.8 billion. When adding in anticipated foreign military sales, orders from NATO countries, and replenishment spares, the total value of supplying engines for just the F–16 could surpass $20 billion. Pratt had come a long way from its first sale of six engines for $90,000 and seemed to be locked in securely for many years of high profit. However, other events were occurring which would greatly change Pratt's outlook. Performance problems were cropping up in both their military and commercial businesses.

Pratt had won the competition to provide engines for the F–111, but by the mid-1960s, the F–111 and the Pratt engine, the TF30, were not meeting design requirements. One of the major aircraft problems was excessive weight, which then required even more thrust from the engines to meet flight performance goals. Because the TF30 was not working properly, unfortunately it could not meet the original requirements, let alone provide any additional capability. Mating the engine to the engine inlet of the airframe was especially complicated. After tests found the engine behavior erratic and the thrust consistently inadequate, several major design changes improved but could not correct all the deficiencies. Moreover, Congress, aware of the problems, reduced the size of the program. As a consequence, Pratt's expected profits evaporated rapidly.[4]

Pratt's problems in its commercial engine business were even more perilous. Aerospace industry designs for the C–5, the huge Air Force airlifter, spawned the idea of a commercial jumbo jetliner capable of carrying several hundred passengers per trip. Pan American saw the potential and took the initiative to arrange for the design and production. Lockheed, the C–5 manufacturer, was uninterested in also taking on the airline business, but Boeing was eager and agreed to organize an industry team to produce what was to become the 747. Pan American insisted on a single contract with Boeing for the airframe and engines plus guarantees that performance stipulations would be fully met. Boeing preferred that Pan American contract separately with the engine manufacturer in order to reduce Boeing's exposure to financial losses from the additional burden of responsibility. Nevertheless, the prospect of getting in on the ground floor with the commercial jumbo jets made the risk acceptable, especially if most of the risk fell to the engine manufacturer. That is exactly what Boeing did.

Just as Lockheed was too preoccupied with making the C–5 to get involved with the airlines, so too was GE, busy providing the engines for the C–5, unwilling to join with Boeing in the 747 venture. However, Pratt agreed and signed a very demanding contract. Pratt guaranteed to deliver engines with sufficient thrust to meet Pan American's performance requirements regardless of the Boeing airframe weight.[5] Pratt's guarantee led to huge financial losses.

Boeing told Pan American at first that the 747 would weigh 550,000 pounds, but the plane gained another 110,000 pounds within a year.[6] Pan American was furious. The added weight meant the airplane had to carry fewer passengers and that profits would be lost unless the engines produced extra thrust to compensate for propulsion of the heavier airframe and made possible the originally planned passenger load. Pan American insisted that Boeing comply fully with the contract. Boeing, in turn, sent the same message to Pratt, who could not pass the responsibility any further. Pratt's losses on the deal with Boeing started piling up in the late 1960s because of continuous re-engineering, higher spare parts consumption, and the need for frequent replacement of engines. Pratt eventually reported total losses of $250 million on the 747 engine and in 1971, for the first time in 37 years, reported to stockholders a financial loss from the year's operations.[7]

United Aircraft's board of directors was, of course, disturbed by these losses. Moreover, the directors were generally dissatisfied with United's recent record of capital investment. So, in 1971 they launched a search for a new chief executive officer and found Harry Gray. Gray had been the number three man at the Litton conglomerate and had developed successful skills in furthering corporate expansion through acquisitions and mergers—precisely the talents United wanted.[8] Gray, hired in September 1971, immediately set out doing what he was hired for, expanding. "His name became a byword for aggressive merger makings."[9] Soon the corporate name was changed from United Aircraft to United Technologies Corporation (UTC) to reflect a widening range of corporate activities, such as Otis Elevator, Essex Industries (a wire manufacturer), and Carrier Corporation.

A key ingredient for the success of UTC expansion attempts was a readily available supply of cash and borrowing power. Notwithstanding the losses on the 747 engine, with Pratt's military engine contracts, UTC had plenty of both.[10] With high corporate priority on maintaining this steady, high-volume inflow of cash to support its corporate expansion policy, UTC could not afford a repeat of the 747 debacle. This helped explain why Pratt was so anxious to escape the correction of deficiencies clause. Pratt could not tolerate costly performance guarantees that might crimp further expansion efforts. In the years ahead, when negotiations over fixing engine problems became increasingly difficult to complete, some people in the Government were to suspect that Pratt's adamant hardline positions and excessively legalistic contract interpretations stemmed from top management's overriding concern to support Harry Gray's cash flow and expansion objectives. But while Pratt struggled, one of its competitors prospered.

GENERAL ELECTRIC

General Electric, started in 1892, moved to the forefront of American technology in generating electric power, particularly through turbines. By 1917, some of GE's pioneering work in turbine technology had caught the attention of Government aviation officials. Soon GE was developing an aircraft engine turbo-supercharger which boosted engine power at high altitudes; however, GE was not alone in this work. Even in its initial involvement with aircraft engines, GE had to compete for business. And it won. After a competitive flight demonstration in 1918, GE's supercharged engine was found superior to its competitors', and GE won its first military aviation contract.[11] From the contract came resources to develop further its technology, and performance advancements followed regularly. For example, when Billy Mitchell set out to prove the effectiveness of aerial bombardment from over 15,000 feet, he had GE superchargers installed on his bombers to reach the needed altitude and range. Mitchell sank the target ships, proved his point, and ensured GE of a significant role in the continuing evolvement of military aviation.

Before and during World War II, GE's work in aircraft propulsion was two-pronged. General Electric turbochargers were in great demand not only by the US military but also by US allies, keeping production at GE manufacturing facilities at high levels. However, there was another ongoing effort by a relatively small group of GE researchers and technicians which led the way to jet engine development. At first on its own initiative, later with encouragement and support from Hap Arnold, GE was working in cooperation with the National Academy of Sciences and the British to harness gas turbine technology for aircraft propulsion.

General Electric was well-suited for this secret work of the early 1940s done to develop a jet engine. The GE turbo-supercharger is much like a small jet engine since both use air compression and turbine drives. So GE understood many of the turbine dynamics inherent in jet engines. Then when GE contracted with the Navy to develop a gas turbine engine for PT boats, the company inadvertently leapfrogged into an even better understanding of how to develop a gas turbine engine for aircraft. In order to construct a combustion chamber and single-stage turbine at its factory, GE had to downsize all the components, for fear that a full-size unit would overwhelm the compressed air capacity of the test facilities. As a result, when the components to power a PT boat were reduced in size, GE was much closer to having a jet engine suitable for an aircraft.[12] After intense work by an Army Air Corps/Bell Aircraft/GE team to refine the technology, on 2 October 1942, a jet engine powered XP–59A flew successfully, and the US jet age began.[13]

After the war, development concentrated on making more powerful, fuel-efficient engines. General Electric directed its main effort toward the military market and in the late 1940s and the 1950s, produced the J47 which powered numerous different military aircraft such as the F–86 and B–47. Later GE developed the engine for the F–4, which is still the largest part of our fighter force. General Electric prospered in this market and expanded its facilities. However, opposite results met the company's work in the commercial market.

While Pratt was teaming with Boeing for the earliest 707 and with Douglas for the DC–8s, GE, rather than being shut out of the airliner business completely, decided to team with General Dynamics for the Convair 880s and 990s. The venture proved to be a big mistake, often described as "the greatest American business failure in recorded history."[14] Although General Dynamics built technically sound aircraft, its competitors held too large a lead in the marketplace. General Dynamics and GE never found enough demand for their product to recapture the costs incurred in development. *Forbes* has estimated the GE loss on the project at about $150 million.[15]

Until the 1970s when airlines started re-engineering to achieve greater fuel economies and to meet tough Government standards for air and noise pollution, General Electric remained almost totally out of the commercial market. When this re-engineering happened, GE was ready to compete with Pratt. GE had teamed with the Societé Nationale d'Étude et de Construction de Moteurs d'Aviation (SNECMA), a French government-owned aerospace company, to produce a turbofan jet engine which some airlines selected to replace the Pratt JT3D. Also to GE's advantage, airframe manufacturers moved away from offering their planes with only one choice of engine design. Several interchangeable turbine configurations were available on given planes at the time of original sale. Pratt was no longer guaranteed that a Boeing sale was synonymous with a Pratt sale. Airlines sometimes selected the Boeing aircraft with a GE engine. In 1966, GE received 1.7 percent of the orders for large commercial aircraft turbines, and Pratt received 92 percent. However, by 1978, under this new practice of separately selecting the engine for each aircraft sale, GE received nearly 25 percent of new orders, while Pratt's share dropped to 63 percent. The rest of the orders went to Rolls Royce and the SNECMA consortium.[16]

Both Pratt and GE learned that success in the aircraft engine business can be short-lived. The outlook for future business can appear very favorable based on contracts received and the prospects for continuing follow-on orders. However, circumstances can change suddenly with profound consequences on the fortunes and direction of the company. At one moment the picture might appear very bleak with few contracts in-hand and little reason to hope for improvement. But in the aircraft engine business, events continually arise to challenge the dominant position of one rival and provide opportunity to the other. A simplified comparison of Pratt and GE illustrates the point nicely.

Coming out of World War II, Pratt was clearly oriented toward the commercial aircraft engine market and GE toward the military. As GE was persistently trying and eventually succeeding in significantly penetrating the commercial market, some of its military work was faltering. On the other hand, while Pratt was losing ground in the commercial market, it was making inroads to the military. General Electric won the contract to supply the C–5 engines. However, the engine development costs exceeded expectations, and the aircraft design and cost problems led Congress to reduce the number of aircraft ordered. With lower volume of production, GE had significantly less opportunity to amortize its costs and took a substantial loss on the project.[17] General Electric did win the contract to develop the engine for the SST and also won the competition to furnish the F101 for the B–1. Unfortunately, congressional enthusiasm for the SST was insufficient for funding production. Then President Carter cancelled the B–1 production program in 1977, and GE no longer had a substantial military business base. Meanwhile, Pratt, although losing ground to GE in the commercial market, was gaining substantially in competing for the military dollars. Pratt developed the engine for the SR–71 and then it won the competition to power the F–14s, F–15s, and F–16s. Though each firm had its ups and downs, overall, in the early 1970s Pratt was the leader in the aircraft engine industry.

However, GE was not idly observing this seesaw movement of its fortunes. In the early 1960s, GE had studied future military and commercial requirements and had decided on a "core" or "building block" concept to meet the broad range of future business opportunities. This concept was fundamental to GE successes 20 years later.

The basic idea of the GE concept was to design a family of engines, each one aimed at a specific market. The family would have an identical, or nearly identical, core consisting of the compressor, combustor, and turbine. Depending on the aircraft to be powered, the combination of additional engine components (fans, afterburners, and thrust vectoring devices) could be tailored to meet the exact performance requirements for the aircraft. With this building block scheme, having the common engine core upon which all other tailor-made components were added, GE could compete for virtually any type of aircraft jet propulsion system, save costs in manufacturing, and save time in meeting schedules. The project for this concept, established in February 1962 and designated GE1, cradled the company's hopes for the future. As reported in GE's official corporate history, the GE1 "building block concept is perhaps the most significant business/technology achievement to date in its aircraft engine history."[18]

One of the movers behind GE1 was Gerhard Neumann, a legend in the world of aircraft engines. Neumann worked 32 years for GE and retired in 1980 as head of the multibillion dollar engine group. His name usually conjures images of top management people-orientation and dirty-fingernail know-how. Neumann was born and raised in Germany where he received training in mechanical engineering and a mechanic's apprenticeship. When World War II started, Neumann was in Hong Kong. His mechanical skills soon led him to General Claire Chennault and the famous Flying Tigers. Neumann, also known as "Herman the German," proved invaluable as Chennault's crew chief. Among his many hands-on experiences, he assembled a Japanese Zero, the first for the Americans, without any drawings or manuals, from the parts of several captured and partially destroyed Zeros.[19] Neumann also completed special missions for General "Wild Bill" Donovan of the Office of Strategic Services. In recognition of this work Donovan helped Neumann obtain US citizenship through a Special Act of Congress in 1945.

Rising through the ranks in GE, Neumann earned a reputation for his knowing the "trenches," for getting out into the shops among the mechanics, talking directly with employees, and stressing quality. Neumann's "walking the shop" philosophy took him out to the flight lines and into the hangars where engine maintenance is a daily affair. Pursuing this philosophy, in 1967 and 1969, he had the unique permission of the Air Force Vice Chief of Staff to travel in Vietnam to seek out the firsthand experiences of Air Force mechanics working on GE engines.[20] At GE plants, he was once known for "Father Neumann's Tent Revival Meetings," at which he would personally exhort thousands of GE employees to accomplish more. Knowing the most fundamental aspects of the business and working closely with the people in the trenches both stood him well and contrasted with Pratt's leaders. For example, Robert Carlson, UT's executive vice president with responsibility over Pratt, hired by Gray in part to turn around Pratt's relationship with the military, had been with a farm equipment manufacturer. Carlson did do much to help improve Pratt's dealings with the Air Force. Yet, perhaps even more could have been accomplished if Carlson and other key UTC executives were better versed in the fundamentals of the aircraft engine business, as was Neumann, and had a stronger image with personnel.

Overall, Neumann exuded a business philosophy that was more in line with the military's perception of treating customers than were Pratt's practices. In his autobiography, Neumann explained his view:

> From close relationships with our customers over many years, I learned one thing: Building personal confidence, listening to what the customer thinks—regardless of what the computer says—keeping my promises and an all-out, stubborn dedication to meet or exceed commitments that I made were the most important reasons for success in my job. An occasional engine failure was forgivable—even expected—but failing to make every effort to fix problems and meet promises was not.[21]

GE PROSPECTS ON THE RISE

Though indicators pointed down for GE's future in military engines after President Carter cancelled the B–1, the company's strengths proved invaluable in taking over leadership in the industry by the mid-1980s. For one, GE's building block concept allowed it to pounce quickly on any new opportunity. General Electric also had a respected reputation for emphasizing quality and service, *a la* Neumann. In addition, although B–1 production stopped, the development work continued, including work on the F101 engine. Moreover, Pratt was having increasingly serious problems with the F100 and apparently finding no solutions. The longer Pratt took to satisfy the military, the better GE's opportunity to develop an alternative to the F100. Thanks to the in-place core concept, GE would not have to start from scratch to demonstrate an alternative.

Some time early in 1975, under Morris Zipkin GE began to develop an engine superior to that in the F–14 and competitive for other possible fighter applications. Some even spoke of the F–15 and the F–16.[22]

Because of the GE1 project, GE instinctively looked around for additional opportunities to tailor the F101 core to specific aircraft requirements. When GE looked at the F–14, Pratt's poor progress convinced GE of a real opportunity to move into the market.[23] Then in fiscal year 1977, Congress made the opportunity irresistible when it authorized and appropriated $15 million to begin development of an alternate engine for the F–14. Though the Navy did nothing with the money, the next year Congress added another $26 million for the same purpose. The Navy again did nothing with the money, since the issue was not how to pay for developing an F100 alternative, but how to afford the enormous cost of subsequently retrofitting the F–14 fleet with a new engine. The Navy estimated the retrofit cost at $1 billion and believed the price was unaffordable in the face of budget limitations and other priorities. That's why the Navy didn't use the money. But the money's mere presence was an encouraging signal to GE, showing public dissatisfaction with Pratt's engine and significant congressional resolve to get something better.

Of course the $41 million didn't suddenly emerge from the Congress without outside influence. GE was very much behind the idea, working effectively, for example, through Congressman Sam Stratton, whose district was in up-state New York, a GE stronghold. Congressmen Dickinson and Whitehurst and key House Armed Services Committee staffers, such as Tony Battista and Tom Cooper, played major parts in working the necessary legislation through Congress. When Cooper became Assistant Secretary of the Air Force in 1982, he had responsibility for Air Force acquisition programs. His background with the engine issues helped to set key policies for the competition between GE and Pratt.

Nor was Pratt oblivious to what was happening in Congress. However, in a style consistent with its approach to other F100 challenges, Pratt chose to attack the competition and the idea of an alternative, rather than focusing its primary energies on fixing its engine.

Meanwhile, in March 1976, before knowing for certain that the Navy would not spend the new money from Congress, GE switched its sights from the F–14 to the F–16. GE felt that even though the Navy needed a replacement for Pratt's underpowered TF30, the Navy would be reluctant to use a derivative of an Air Force engine.[24] General Electric well grasped the realities of inter-Service rivalry.

Events fell further into place in November 1977, when Neumann visited General Dixon, Commander of the Air Force Tactical Air Command (TAC). At TAC, the general faced the serious F100 problems of unsatisfactory maintainability, reliability, operability, and durability. The prospect of worsening problems with the single-engine F–16 loomed large in TAC's concerns. Neumann, the consummate aircraft engine guru with longstanding, respected relationships with Air Force senior leadership, mentioned that GE was working on an F100 alternative. Shortly after the visit, Dixon sent his message to General Slay at Air Force Systems Command, emphasizing the need for an alternative to sole reliance on the Pratt engine.

SUBCONTRACTOR STRIKES

If problems with the engine's performance and the perception of Pratt's management attitude were not alone sufficient to justify an alternative to the F100, the impact of two strikes by Pratt's key suppliers made up the minds of all but the most die-hard Pratt loyalists. In April 1979, Ladish, a forgings subcontractor, and Fafnir, a bearing manufacturer, were each hit by strikes. Both subcontractors saw labor walk out for five months until new contracts could be ratified.

The impact of the strikes overwhelmed the Air Force. It did not even seem to matter that Pratt had not caused the strikes. What mattered? The Pratt engine was again the source of acute problems for the Air Force. The strikes re-emphasized the precarious nature of F–15 and F–16 total dependence on one manufacturer and one engine to meet all the high-performance propulsion needs of the tactical air forces. During the worst periods of shortage in 1980, the Air Force was 90 to 100 engines short.[25] Engines were in such short supply that two F–16s undergoing tests rotated flight time in order to share the same engine. One pilot quipped to a Pratt representative, ''I see you have a spare airplane for your engine.''[26]

Navy estimated the cost of retrofitting the F–14 at $1 billion. Shown, an F–14A Tomcat after taking off from the USS America. *(An official US Navy photo, photographer CWO Joe Leo.)*

The Air Force devised several elaborate plans to mitigate the effects of the strike. For example, F–15s undergoing modification had their engines removed. Once removed, the engines were sent to the McDonnell Douglas factory in St. Louis where they were installed in F–15s just coming off the production line and used to fly the planes to an Air Force base instead of leaving the planes at the factory where they would incur expensive storage costs. At one point, the crisis became so acute that the Air Force even used impounded Iranian engine parts to meet critical US Air Force requirements.[27] In addition, the Air Force established a system of priorities with its first obligation to support the existing F–15 and F–16 force structure. After that, available supplies went to new aircraft production and foreign sales. The priorities and allocation procedures of the Defense Production Act of 1950, invoked through the Department of Commerce, helped to account for and allocate all subcontractor and Pratt F100 materials to satisfy priorities. General Slay estimated that Commerce's assistance under the provisions of its Defense Production Act saved about six months in the strike recovery.[28]

Even considering all the elaborate plans, extra work in shifting engines around, the missed training opportunities, the additional costs, and the greater vulnerability of national security, as their most enduring impact, the strikes' dramatic and costly consequences coalesced opinion among Air Force general officer leadership. In spite of the high price tag on developing an alternate engine, as well as the opportunity cost to forgo other needed programs under a constraining budget to pay for the alternate engine, the program to get an alternate began to look affordable under the circumstances. Because of the effects of total dependence on one engine, the Air Force resolved to get a second source. The accumulation of F100 problems,

Pratt attitudes, and the strikes jelled united commitment to bring along the GE engine.

5
THE BIGGEST BATTLE OF ALL

On the previous page, the F–16 frontline fighter in a vertical climb demonstrates dramatically the technical achievements accomplished in jet engines. (An official US Air Force photo, photographer unknown.)

THE MAJOR OBSTACLE to developing a General Electric F101 derivative for fighter aircraft was money. In 1978, as a few staff people at headquarters of Air Force Systems Command and at the Air Staff searched for ways to fund GE, their attempts were repeatedly rejected by the consensus of the senior Air Force leadership, who liked the idea of a derivative but not the budget sacrifice it would entail.

Existing congressional budget authorizations and spending ceilings on the Air Force had no room to start a development program aimed first at quickly demonstrating the technology through a series of flight tests and ultimately at manufacturing thousands of engines for the tactical fighter fleet. Reprogramming the funds from Air Force projects already approved within the ceiling appeared to be the only alternative. Relatively little was needed, initially just $30–40 million to supplement the work GE had already started. However, for a long time even this amount was inaccessible. Until the subcontractors' strikes of 1977 provided impetus, the derivative or alternate engine program could not muster adequate support within the Air Force hierarchy to take money away from already established Air Force programs. Consequently, those who remained undaunted in their belief in an alternative to the F100 needed to think more creatively of ways to get the money.

THE ENGINE MODEL DERIVATIVE PROGRAM

In the 1960s, Congress saw misuse of the Component Improvement Program (CIP). Intended to correct engine performance deficiencies revealed through operational use, the program, in a way, developed new, follow-on engines. Congress opposed development of new engines in this way because the procedure circumvented its oversight of formal Department of Defense budget requests for research and development projects. Developing some new engines under the guise of component improvement to existing engines shielded that development work from the normal congressional staff scrutiny and inquiry in open hearings which examined military development projects.

Although each project in the military research and development budget is separately identified and accounted for, the CIP annual budget is a lump sum. Individual CIP projects are not normally examined because in theory no one knows for sure what engineering projects might need quick reaction during the year to correct problems discovered in an engine's performance. Annoyed with the obfuscation of developing under the guise of improving, Congress stipulated new rules in 1968 to prohibit such development work under CIP. Did Congress overact? Some engineers believed the rules went too far and stymied some legitimate, much needed, engine improvement which might have evolved into engine derivatives. Congress, however, thought it had provided for continuing development of derivatives by also creating a new, separately identified, budget line—the Engine Model Derivative Program (EMDP).[1]

To Congress these strict new rules only changed the budget line under which derivatives would be identified. But the effect was much more profound. Forcing development of derivatives into a separate budget line also forces each such project to compete for approval within the military services. Just as Congress would have visibility and right of approval of proposed projects, so too would the layers of budget review within each of the military services. Although some budget reviewers might be unwilling or unable to investigate the detailed rationale behind a proposed CIP funding level, the segregated EMDP work was an easy target for inquiry. Accordingly, it was very difficult to get EMDP funding approved. Nearly 10 years later, Lieutenant Colonel Phil Hughes, responsible at the Air Staff (later at Headquarters, Air Force Systems Command) for EMDP projects, was unable to get just $1.9 million approved for EMDP.[2] In general, EMDP funding was essentially unused; under it derivatives were not developed.

In 1978, however, because circumstances changed, EMDP became the budget mechanism for funding GE's continuing work to develop an F101 derivative for fighter application. General Dixon's message to General Slay lamenting the vulnerability of a single engine and source of supply was the catalyst for funding development of an alternate fighter engine under EMDP.[3] Nevertheless, even after Air Force Systems Command decided on the program under which it wanted to develop an F101 derivative (F101X), it still lacked the money. Air Force development needs were too great for available funds, and the views of Air Force senior leadership on the engine problems were still too fragmented to yield a favorable consensus for the F101X. Arranging for the money would require a creative, hard fought, well-coordinated effort.

While Phil Hughes worked on the problem for General Slay at Systems Command, Dean Gissendanner, an Air Force civilian employee, tackled the same issue on the Air Staff. Hughes and Gissendanner, in harmony on the importance of getting the F101X project funded under EMDP, sought to overcome seemingly endless barriers to success. Each knew propulsion technology, was well "networked" in the propulsion community, and had superb staff skills. Hughes, for example, had spent several years in the R&D propulsion office at Wright-Patterson Air Force Base and just prior had been responsible for engine projects on the Air Staff. Hughes and Gissendanner knew their business. In addition, as this core group got into the full swing of efforts "to sell" the program, Majors Bill Eddy, Riley Shelnutt, and Tack Nix, also staff-smart and well-motivated, ably assisted to make their efforts succeed.

At the same time, Nick Constantine, General Electric's contribution in Washington to Air Force efforts to find money, touted a style somewhat akin to Gerhard Neumann's. He, too, emphasized service to the customer and communication with the people in the trenches. Constantine did most of his work with majors, lieutenant colonels, and their civilian counterparts, the people involved with the daily rigors of persuading, explaining, coordinating, and evaluating all aspects of Air Force propulsion programs in the Washington, DC, area. No one knows who first thought of the Navy money, but somehow the dedicated work of these insiders uncovered the Navy's unused $41 million from Congress. While Hughes and Gissendanner briefed Air Force generals on this possible source of funds, Constantine used GE contacts on Capitol Hill to check congressional receptivity to an Air Force raid on the languishing $41 million. The word back from Congress was encouraging and came primarily from Tony Battista on the professional staff of the House Armed Services Committee.[4] Battista, of course, could provide no guarantees. However, Congress, disturbed by the Navy's continued failure to use the money made available to correct the F–14's engine problems, knew of the burgeoning F100 problems and grew increasingly concerned. Generally, Congress seemed favorable to the Air Force using the money. Accordingly, General Slay called his Navy counterpart to let him know of the Air Force intent and to establish a dialogue for arranging the mechanics to make the funds transfer possible.[5]

At the outset the objective was only to have an alternative to sole reliance on the F100 for 100 percent of the propulsion for the frontline tactical air forces. Although Air Force officials close to the problem hoped that funding development of the F101X would also prompt better responsiveness and performance by Pratt, no one seriously entertained the notion of actually competing Pratt against GE for future requirements. The idea was simply to have a stand-by, demonstrated technology.[6]

Meanwhile, as efforts continued in arranging funding for the F101 derivative under EMDP, high-level, regular discussions with Pratt did not always succeed in even getting the parties together. General Slay, often in the forefront of the interaction, felt his frustration and annoyance

General Alton D. Slay, as Commander, Air Force Systems Command, was deeply involved in funding the F101 derivative.

grow ever larger as Pratt seemed incapable or unwilling to satisfy Air Force requirements. At one point, Pratt marketeers visited General Slay to talk about the possibility of the Air Force buying a more powerful F100. General Slay dismissed them with the admonition that Pratt had no business talking about growth when it could not even fix the problems with the current engine.[7] The problems did not involve a need for more power. The Air Force wanted Pratt to make the existing engine more reliable, maintainable, and operable; for this purpose hundreds of millions of dollars annually went to Pratt through the Component Improvement Program.

On another occasion, General Slay called the Air Force Chief of Staff, General Allen, and asked him for a meeting in his office with Harry Gray. General Slay felt that he had reached the end of his rope.[8] At the meeting, General Slay recounted the details of his problems with Pratt's performance, focusing on problems with the engine and Pratt's lack of management responsiveness. General Allen backed up General Slay's portrayal of the problem's seriousness. This meeting may have been the first occasion Gray really sensed the depth of Air Force concerns and frustration.

Gray met with General Allen again in October 1978 at Hill Air Force Base, Utah, where the ''Chief'' was holding one of his regularly scheduled meetings with his commanders and senior staff to consider topics of high interest. Gray briefed the Air Force on Pratt's recovery efforts and pledged the total resources of the corporation to the success of the F100 program.[9] Pratt believes that at this meeting Gray secured a pledge that if the company lived up to its promised performance, the Air Force would not proceed with development of an alternate engine.[10] This interpretation of the meeting is important, possibly accounting for some of Gray's subsequent personal contrariness when he saw the Air Force moving forward with the alternate engine toward a competition with Pratt, in spite of what he considered substantial performance improvement by the company.

In the meantime, back at the battle over EMDP funding, Air Force, Navy, and OSD staffers worked on a strategy to improve fighter engines. The "Joint Conference Report on the Fiscal Year 1979 Appropriation Bill" directed the Air Force and Navy to undertake a joint competitive program to develop an alternate engine for the F–14, F–16, and other aircraft. Responding to Congress, the Services initiated a three-point strategy aimed at fixing the immediate problem as well as providing a solid foundation for future engine development.

Under the program, crafted to harness competitive pressures to assure its success, CIP funding continued to underwrite Pratt's work to improve the F100 and the TF30 (the F–14 engine). Second, EMDP funds would be used primarily to supplement GE's initial work on the F101X, which had been started with company money, and also to provide a modest amount of EMDP money to fund Pratt in developing a higher thrust F100 variant. However, the primary objective of the military's dealings with Pratt continued unquestionably to be correction of the current engine problems. Third, the Services agreed to fund a study of new engine requirements into the 1990s.

After much back-and-forth discussion of conditions and the mechanics of transferring funds from the Navy to the Air Force, often requiring GE prodding and general or flag officer interaction on the particulars, the Services signed a Memorandum of Understanding (MOU) in February 1979, titled the "Memorandum of Understanding Between the USAF and the USN for Cooperative Aircraft Engine Demonstrations and Development." General Allen signed the document on 3 February and Admiral Haywood on 12 February. The sum of $41 million was split between the Services, giving $33 million to the Air Force and $8 million to the Navy. Soon thereafter, both GE and Pratt began getting research and development funds to demonstrate better engines. Available but left essentially dormant for over a decade, EMDP was finally used in a substantive way to launch an unprecedented effort to reintroduce competitive pressures into a major acquisition program, nine years after Pratt had seemingly locked up that market.

On 5 March 1979, the Air Force contracted with General Electric for initial development of an F101 derivative for fighter application. GE's task was to put the engine in the F–16 and fly it successfully before 30 months had elapsed. At a lower level of effort, Pratt was funded to demonstrate, only through ground tests, a higher thrust F100.[11] The split of funds between GE and Pratt was about 90–10. Mindful of General Dixon's earlier communiques, expressing his alarm at over-dependence on the troublesome F100, General Slay sent a letter to Dixon in March 1979 reporting that the EMDP had been "structured to provide propulsion options for our fighter force in the 1980s."[12] If successful, the program provided an alternative to sole reliance on the F100.

SKIRMISHES IN CONGRESS

The common picture of Pratt's reaction to the increasingly clear threat of resurrected competition from GE is stubborn, often acrimonious resistance. The Air Force strategy made sense. Although the F100 was improving, it was still unreliable, too costly to operate and maintain, and too crucial to the effectiveness of the total tactical air fighting forces. Moreover, in spite of years of prodding, Pratt management had been extremely difficult to work with. Sometimes it seemed that, at the highest management levels, Pratt was more interested in the commercial market and other profit-generating ventures. Then, too, Pratt had annoyed many key Government officials. As Congressman Dickinson expressed it, "Pratt took the attitude, 'Like it or lump it. You got no place to go anyhow.' "[13] Nurturing a potentially competitive alternative seemed to be a smart move by the Air Force.

However, Pratt took a different view. When looking at the F100 program from its inception, Pratt was right in arguing the Air Force got what it asked for. The F100 was a superb propulsion system when it worked. All the concerns about durability to withstand thermal cycles, for example, were unspecified in the original F100 contract. Furthermore, incessant pressure to meet the IOC allowed little or no time to perfect the engine. Strictly speaking, Pratt was justified to some extent in holding out for more money before acting on various improvement projects. Moreover, Pratt progressed on F100 improvements as Gray had promised the Air Force. But the Air Force was not adhering to the agreement Gray thought he had with General Allen—the "promise" that Pratt's progress would cancel the Air Force move to develop the GE alternative. Consequently, Pratt felt not only normal business reactions to competition, but also personal irritation at the highest management levels. Whether justified or not, Gray was annoyed.[14] Perhaps Pratt's ultra-strong resistance to competition reflected Gray's personal irritation. Certainly animosity existed on both sides.

While GE was usually working behind the scenes through the efforts of Nick Constantine and other GE facilitators to help its cause, Pratt was using every resource available to stop the Air Force work with GE. Pratt attacked the concept of a competitive alternative and the manner in which the Air Force was structuring the EMDP. Eventually Pratt would label the competition "unusual, illogical, unique and expensive."[15] For its part, the Air Force thought Pratt should have devoted more energy to fixing the engine, thereby showing that an alternate was unnecessary.

Because the availability of money was the key element for success of the Air Force strategy, Pratt chose to fight the EMDP at the source of the money, Capitol Hill. In February and March of 1979, a series of thorough, intensely debated congressional hearings were held on Air Force intentions toward the F100 and a GE alternative. Leading the charge for Pratt was the Connecticut delegation, particularly Congressman Giamo, in hearings that became a forum for venting all of Pratt's concerns.

Regarding the very idea of an alternate engine, for example, Congressman Giamo said, "the concept of a standby fighter engine I think is something relatively novel and certainly sounds like an awful lot of money that could possibly be expended elsewhere."[16] The allocation of money to GE and Pratt was also challenged. Another congressman said about the EMDP, "I am amazed a dollar split of 90–10 is considered a competitive approach."[17] In early February 1979, Senator Weicker, a Republican from Connecticut and a member of the Senate Appropriations

Committee, sent a letter to Dr. Perry in an attempt to thwart the letter contract to GE. The senator noted that Congress wanted the alternate fighter engine conducted on a competitive basis while the Air Force headed in a sole-source direction. Senator Weicker asked that "further action on this contract be suspended until it can be thoroughly discussed in the impending DOD Appropriation hearings."[18]

General Allen was very candid in his responses to the questions from Congressmen about the need for an alternate fighter engine:

> We are anxious that Pratt and Whitney sees this situation as one in which their interests are best served by improving the durability and maintainability of the engine at acceptable costs.... *The only issue at hand is what happens if they fail or are insufficiently motivated....*[19]

About the 90–10 split General Allen explained:

> Pratt and Whitney is receiving a very large amount of money for the production of the F100 engines. Associated with that, it is receiving funds in the Component Improvement Program.... The primary purpose of the development program ... is to bring General Electric ... into a position of being able to *provide us some alternative and some of the advantages of competition* in the area of the F100 engine.[20]

In reply to a similar question about the fairness of the 90–10 split, another Air Force general, appearing before the Senate Armed Services Committee, stated:

> To obligate equal EMDP funds to Pratt and Whitney and General Electric would further contribute to an *imbalance in the competitive situation which currently favors Pratt and Whitney....* Pratt will be working from a development funding base on the F100 and TF30 engines in excess of $2 billion and an experience base in afterburning turbofan

fighter engines of 14 years. GE will be working from a development funding base on the F101 of $0.7 billion with no previous experience.[21]

Dr. Perry did appear in hearings before the Senate Appropriations Committee in late February in 1979 as Senator Weicker had requested, but the Air Force plan was unaffected by the hearing, and the letter contract was issued to GE in early March 1979.

So, the EMDP was launched, even in the face of a resistance that wouldn't go away. However, funding for the full term of the program was not assured. The three-year program was short $14 million in fiscal year 1980, and no financial provision existed to continue the program beyond completion of flight demonstrations in 1981. Nevertheless, at least the program had started. Repeated trips to Congress would be necessary to get more money, and each time the maneuvering would be tricky with substantial perseverance and negotiating needed for the program to survive.

For instance, one of Pratt's supporters in Congress once made an unexpected move in a subcommittee of the House Armed Services Committee to transfer $25 million budgeted for GE's work to Pratt for work on increasing F100 thrust. The surprise motion in the subcommittee caught GE supporters unprepared, and Pratt's interests prevailed. However, in the next legislative step, presenting the same proposal before the full committee, Congressman Dickinson succeeded in restoring the authorization to the original plan. The congressman explained that "I sort of got blind-sided in subcommittee because I didn't know who was working this thing behind the scenes.... It sort of got to be a personality thing."[22]

The effects of the subcontractor strikes, beginning in April 1979, did much to dispel doubts and criticism of the

EMDP. Turmoil in Air Force logistics and operational capabilities united Air Force support of an alternate fighter engine as events of less traumatic magnitude probably could not have. The strikes became another one of those unforeseen circumstances that propitiously moved the alternate engine past obstacles perhaps otherwise insurmountable.

Nevertheless, resistance to the program never went away completely. General Electric and Pratt continually jockeyed to exploit advantages and avoid pitfalls. For example, in the fall of 1979, the Senate Armed Services Committee held a hearing on the F–15 and F–16 engine problems. People close to the situation believe GE arranged for the hearing through the committee staff to focus attention on the F100 problems and also on the need for funding an alternate engine. At that time GE and some Air Force staffers were seeking about $34 million in additional funding for fiscal year 1981. General Slay was asked specifically to testify, but he didn't welcome the opportunity.[23] The Air Force preferred to avoid the publicity and fanfare of a hearing solely on this topic. Better from its vantage point were small meetings, quiet development efforts, and diffusing the appearance of a GE versus Pratt confrontation. The Air Force did not want to force Congress to choose sides between the contractors. Instead, Air Force strategists hoped to stick to objective discussion of the engine problems in low-key fashion, keeping politics out of the debate as much as possible.

Nevertheless, once called upon to testify, General Slay prepared in exhaustive detail, with hundreds of viewgraphs and exhibits to explain the dimensions of the engine problems thoroughly and understandably. Experts from the propulsion office at Wright-Patterson Air Force Base spent long hours over many days providing General Slay with fact after fact. The Ohio residents flew home for Thanksgiving, but then right back to Washington.[24] At the hearing, General Slay dazzled listeners with his command of detailed facts and the clarity of his presentation. Adroitly, the general presented a convincing case that the Air Force had a tremendously difficult problem, that the Air Force was doing everything conceivable to help itself with available resources, but that it needed more help to meet minimum standards of operational dependability. In a balanced presentation, General Slay pointed out, for example, that Pratt had done much to improve its production capacity by purchasing a machining facility in Maine, buying $79 million of additional equipment, and increasing the production labor force by almost 6,000.[25] As 1979 drew to a close, Congress was left with the image of competence and objectivity in the Air Force management of engine programs.

However, uncertainty over where the EMDP was ultimately headed continued in 1980. Would the work to develop an alternate fighter engine stop at the end of the EMDP or would full-scale development continue at that point leading to a production program? Obvious repercussions from the strikes supported plans to continue work on the F101 Derivative Fighter Engine (DFE) even after EMDP was completed in the following year. Aware of the trend, Pratt was finally displaying elements of a long-hidden competitive spirit. Talented people hired from the outside filled key executive positions. For instance, Maury Zipkin, who had been in charge of GE's initial work at Evendale to make an F101 variant for fighter application, took over Pratt's advanced programs. Zipkin quickly set out to develop a family of engines somewhat along the lines of the GE1 building blocks concept. In addition, he changed some of Pratt's policies for investing in independent research and development so that fewer experimental prototypes of a particular engine under development would be built for test.[26] Instead of building five or six sets of hardware, Zipkin believed that perhaps one and a half would suffice. In this way, a greater number of different designs could be tested with relatively fixed available funding.

Harry Gray had also hired Robert Carlson away from a senior executive position at Deere and Company to become president of Pratt.[27] There may even have been a hint that Carlson would be groomed to be Gray's successor. In any case, Carlson pleased the Air Force general officer leadership with his willingness to come to them and discuss the main issues.[28] Carlson was accessible, and he promised to improve Pratt's performance, which, of course, was exactly what the Air Force wanted to hear. For example, Carlson visited the Oklahoma City Air Logistic Center repair depot, saw the high delinquency rate for parts due from Pratt, and promised to get the rate down. Carlson also met and corresponded regularly with Lieutenant General Skantze, commander of all the program offices at the Aeronautical Systems Division at Wright-Patterson, to resolve the continuing problems arising from the strikes. Carlson was encouraging, and the Air Force confidence in his leadership and commitment increased.

Confidence-building stories of promised improvement in Pratt's responsiveness and rumors of improvements to the F100 circulated in Washington. Perhaps as a consequence of favorable talk, Dr. Mark, Secretary of the Air Force, and General Allen testified to Congress that support of the F101 DFE beyond flight demonstration would not be required. However, to some senior Air Force officials, it was still premature to consider ending the DFE. The Air Force had received promises from Pratt, but it still needed more engines and parts. Continued dependence on this one contractor kept many officers close to the problem unsettled. Accordingly, General Slay discussed strategy options with General Poe, commander of the Air Force Logistics Command. Shortly afterwards, General Poe, in a personal letter to General Allen, a West Point classmate, reviewed Pratt's poor record. General Poe concluded by noting that the House Appropriations Committee mark-up of the fiscal year 1981 DOD budget included $74 million for continuation of F101 DFE development and by stating, "I strongly recommend we continue this development to the maximum extent possible to provide competitive alternatives in the future should Pratt and Whitney's performance continue to degrade."[29] Pratt was not fulfilling all of its promises, and the Air Force did not want to let the company off the hook.

On the Hill, Senator Chiles of Florida (Pratt's Government Products Division is in West Palm Beach) urged Senator Stennis, Chairman of the Subcommittee on Defense Appropriations, to delete F101X funding in the next budget. Chiles contended the F101 derivative was essentially a new higher thrust engine which would be a "successor, not an alternative to the current F100 engine.... It seems to me that to continue to put funds into the F101X program amounts to developing a new system—a higher thrust engine on a noncompetitive basis."[30] Chiles was stating an often repeated Pratt theme of unfairness based on the disparity between the thrust of the GE and Pratt engines. Senator Cannon and others gave Senator Stennis the same rationale for deleting the F101X (DFE) money. However, on 18 November, General Allen, agreeing with the recommendations of his senior commanders, went on record with Senator Stennis to state the official Air Force position:

> To date the Air Force had not formally endorsed the addition of these funds. However, the success of the F101 DFE

> as a viable engine alternative has resulted in reassessment of increasing the F101 DFE level of effort. The *Air Force endorses* the addition of $25 million to the *F101 DFE program in FY 81*. I have directed that we *continue development efforts in FY 82* so that we have competitive engine alternatives for the future.[31]

That extra $25 million brought the total to $74.1 million for the year. General Poe, in conjunction with General Slay, had successfully convinced the "Chief" to support the F101 DFE. Impressive results from ongoing tests of the engine had helped. Of course, impressive results should have been expected. Not only did GE have an incentive to do well, but also it had had years of relatively low-pressure development schedules to perfect its product.

However, the battle continued. In December 1980, Congressman Addabbo, the powerful chairman of the House Appropriations Subcommittee on Defense, wrote a strong letter to Dr. Mark insisting that the $74.1 million be allocated to the contractors in a competitive manner. Citing the Appropriations Conference Report, Addabbo wrote:

> It has come to my attention that the Air Force proposes to expend fiscal year 1981 funds provided for the Engine Model Derivative Program (EMDP) in a way which does not appear to conform with the intent of Congress. . . . The Air Force should take immediate steps to ensure that all funds made available for the EMDP for fiscal year 1981 are used in such a way that full and fair competition is affected.[32]

Having lost the skirmish to delete EMDP funds, Pratt was now fighting for a larger portion of the program. Pratt, interpreting congressional guidance for the program very narrowly, thought that Congress intended EMDP funds be distributed to contractors on a competitive basis.

This interpretation, of course, most favored Pratt. However, the Air Force and GE interpreted the guidance much more broadly; the USAF knew the real intent because it had set up the program. The objective was to establish competition in the high-thrust afterburning fighter engine business. Consequently, most of the funds would go to GE since that company had received far, far less money from the Government over the last 10 years to develop the technology. Although Pratt was getting less out of EMDP, the company had received hundreds of millions of dollars for R&D and CIP, and everyone close to the CIP understood there would be spinoffs from fixing problems applicable to upgrading performance. Accordingly, in January 1981, an Air Force brigadier general replied to Congressman Addabbo that "... competition in the important area of high thrust, afterburning turbofan engines does not now exist and cannot exist in the future until an engine such as the F101 DFE is developed to the point that it is a viable alternative."[33]

Again and again, and even to this day, Pratt emphasized the matter of thrust disparity between the F100 and the F101 DFE. Perhaps based on experience in F100 development, when the Air Force was insisting on maximizing the thrust-to-weight ratio as a priority over any other capability, such as durability, Pratt could not, or would not, accept that the Air Force was really serious now about making thrust a lower priority than reliability, operability, and maintainability. In Pratt's thinking, thrust was the *sine qua non* of an engine. The company seemed to believe the Air Force emphasis on cost of ownership considerations was either a ruse or bad judgment.

Knowing that military aircraft tend to get heavier over time as additional equipment and payloads are added on, Pratt expected the Air Force to seek greater thrust for its fighters. Because the F101 DFE thrust was about 28,000 pounds whereas the F100 was about 24,000 pounds, Pratt concluded the Air Force was really aiming at getting a higher thrust engine from GE while scheming to exclude Pratt. From the company's perspective, the Air Force's refusal to give it funds comparable to GE's portion under the EMDP was prejudicial.

In response to these charges, the Air Force steadfastly insisted that thrust was not an issue. The Air Force was legitimately and overridingly disturbed by the enormous cost

The F101 turbofan, focus of the EMDP effort and used in the B–1, underwent 1975 high-altitude cell test concentrating on afterburner operation and mid-air starting at the Arnold Engineering Development Center. (An official US Air Force photo, photographer unknown.)

of owning the F100 and by the excessive combat-readiness vulnerability from dependence on a single source of supply for the propulsion of frontline fighters. Unquestionably, too, the Air Force had long, unpleasant memories of its relationship with Pratt. Colonels, such as Richard Steere, who had struggled with Pratt's aloofness during the worst of times, now held positions of increased responsibility. General officers close to the issues, such as General Skantze, were also moving upward. Senior officers took with them strong motivations for pushing competition to ensure corporate responsiveness.[34] They were not driven by personal grudges but by commitment to keep the Air Force out of business situations in which the Service was in the power of a sole-source contractor.

A particularly harsh memory to many senior Air Force officers was the episode with Pratt over putting the Digital Electronic Engine Control into the F100 to improve the engine's safety and operability. Because the Air Force didn't have enough money for the project, some generals struck a deal with Pratt, or thought they had, to split the costs. The Air Force share was supposed to be about $40 million, but Pratt reneged on the deal after the Air Force had struggled to get the $40 million from Congress, telling the Air Force the project would really cost $100 million.[35] The generals were livid. A needed improvement feature was lost. Thereafter, the Air Force never budged on its objectives and sought improved reliability, maintainability, durability, operability, and responsiveness. If in attaining these capabilities, the Air Force also got more thrust, so be it. However, thrust was not a primary, driving consideration.

General Electric saw the Air Force position clearly. From the outset of their program to re-enter the military high-performance engine business in a big way, GE built its marketing strategy around the theme of superior cost-of-ownership. General Electric deliberately avoided performance comparisons, such as thrust.[36]

GETTING THE FORMAL COMPETITION STARTED

In 1981, the United States inaugurated Ronald Reagan as President. As one of his first defense initiatives, the President resurrected the B–1. One hundred B–1B's, with four engines per plane plus spares for a total of 469 engines, went under GE contract. Here, again, an unanticipated event came on the scene. The bomber program provided important benefits to the F101 DFE. For example, because the core of the GE fighter and bomber engines was the same, depot repair capability built for the bomber engines could accommodate the F101 DFE at relatively modest additional cost. In addition, GE overhead costs spread over a larger business base; therefore, the costs of a fighter engine production program would be less. Consequently, the hurdle for selling Congress the idea of establishing a second source was lowered when the upfront costs decreased, thanks to the B–1.

The Air Force also had by 1981 substantial data from the F101 DFE successful flight tests in the F–16. The alternate fighter engine became more than an esoteric notion. Prototypes were available and tested. In fact, the test results were so encouraging that at one point the program office even considered classifying the test data to prevent an uncontrollable stampede for the engine.[37] Thus, in the first half of 1981, the attraction to the GE engine was growing.

Though the obstacles to getting a second source were diminishing, they were far from eliminated. One problem

General Electric won the contract for the B–1 bomber, resurrected in the Reagan administration. (An official US Air Force photo, T.Sgt. Belcher, photographer.)

was how to fund continuing GE work on the engine after the completion of flight demonstrations under EMDP while the Air Force was still deciding when and how to enter full-scale development. Searching for a novel funding mechanism, the creativity and perseverance of the determined core of Pentagon staffers came to the rescue again. A program element called "transition to development," devised and subsequently approved by the Air Force Council, kept GE working on the engine until the Air Force was sure it could proceed with a full-scale program.[38]

Despite the confluence of circumstances which seemed to demand a formal, outright competition between GE and Pratt for future Air Force fighter engine sales, uncertainty over going that far still understandably prevailed. What began as just a program to keep F100 problems from jeopardizing national defense became a program for the

Air Force future. However, doubts continued. No one knew all the costs involved in establishing a second source. Furthermore, Pratt seemed to be improving. Then came another eye-opening event.

In August 1981, the propulsion program office issued a Request for Information (RFI) to industry to gather on a nonbinding and strictly voluntary basis information about prices, product capabilities, schedules, and other pertinent terms and conditions. An RFI sort of says, "We are interested in this product. What might we expect from you, generally, if we were formally to ask for your competitive proposal at a later date?" The GE and Pratt replies were exciting; they provided strong support for proceeding with a formal competition. The industry inputs were irresistibly tantalizing. By the end of the year the Air Force, with General Skantze at the forefront, decided to compete. Even if circumstances were to change and make the competition infeasible (for example, if costs escalated unexpectedly), the competition could still be cancelled with nothing lost.[39]

The acquisition strategy was to harness the forces of competition to get a better deal for the Air Force. There

General Lawrence A. Skantze, then a lieutenant general and commander of all program offices at Aeronautical Systems Division at Wright-Patterson, participated in the 1981 decision to proceed with formal competition.

was no preconceived winner or ratio for splitting the winnings. The Air Force only knew from the replies to the RFI what it had intuitively felt for years, that competition would restore fighter engine capabilities under favorable business terms.

In 1982 the contractors continued work to improve the capabilities of their engines, and based on indications of the Air Force's preferred business arrangements expressed in the RFI, they studied contract terms and conditions they might offer. Meanwhile, the Air Force consolidated its position in support of the formal competition through briefings and one-on-one meetings throughout Washington. In November and December the proposed business strategy was formally reviewed at the Pentagon, and finally on 15 December a draft Request for Proposal (RFP) was sent to the contractors.

The draft RFP, an advance copy of a formal call for industry proposals, completely states requirements. Industry gets it in draft form so that problems with the document can be cleared without the difficulties of formal proposal development. Just as important, however, the draft, not likely to be substantially different from the later official version, gives contractors a headstart toward honing their proposals. The draft saves time. Exceedingly complicated RFPs, particularly for products such as engines and attendant services, have many interrelated special provisions which take time to understand and respond to. However, if the final RFP does not differ too much from the draft, a contractor's proposal can be reasonably requested in just a few months.

SPARE PARTS PROBLEMS

Also during this time in 1982, Pratt was coming under increasing scrutiny for its pricing policies in spare parts. Back in 1977, a cost analyst for the Defense Logistics Agency had reported on "unreasonable results" from Pratt's parts pricing. The Government was paying too much. One solution suggested more competitive bidding for spare parts requirements. Then in 1982, Robert Hancock, a contracting officer at the Oklahoma City Air Logistics Center, found 34 engine parts whose prices had quadrupled in two years. Hancock suspected this was just the tip of the iceberg.[40] He was right.

What became evident to the general public, and especially to the Washington political community, was that spare parts meant big business. In 1982 Pratt sold to DOD more than $600 million in parts.[41] However, as individuals and audit organizations within the Defense Department unraveled the tangle of pricing procedures and contracting techniques for buying these parts, they concluded the military was not getting its money's worth from these enormous expenditures. For example, they found that some of the problem of overly expensive parts stemmed from the contractor's control of information about the items, such as technical descriptive data and sources of supply.

Pratt only made about 20 percent of the F100 parts.[42] For the rest, purchased from outside suppliers, Pratt added overhead and profit margins to the purchase cost when the items were, in turn, sold to the Government. Obviously the parts could be purchased less expensively if the Government went directly to the manufacturers and avoided paying Pratt's margins. However, practicalities complicated matters.

First, an engine is literally thousands of parts. Without a prime contractor (in this case, Pratt) reviewing each of these items to determine the quantities needed, the sources of supply, and reasonable prices, Government logisticians and procurement technicians must perform these tasks above and beyond a normal workload. Hiring more Government people to handle the increased workload offsets some of the anticipated savings from buying parts directly from manufacturers. Second, in order for the Government to go directly to the suppliers for parts, the Government needs the data the prime contractor uses to order parts. These "reprocurement data" are especially important if the Government wants to get the best possible prices. However, prime contractors frequently consider reprocurement data proprietary and unavailable for sale, except at exorbitant prices. Contractors often liberally label their data "proprietary" and leave it to the Government to prove otherwise.

To disprove contractor claims takes not a simple clerk but a skillful technician. The problems expand to getting the data, investigating contractor claims of proprietary rights, and assuring the adequate data for competition. Frequent parts changes, because of improvements to engine design, further complicate the task. The paperwork can be overwhelming. Although more attention focused on spare parts and data problems in 1982, the solutions still seemed either uncertain or too expensive. In any case, no drastic changes were likely at first, although no one in Government was satisfied with the way things were. Pratt, meanwhile, continued to make money from its spare parts business.

THE REQUEST FOR PROPOSAL

Responses to the draft RFP began coming back by January 1983. In addition, before the final RFP was released, at two separate meetings, the Air Force answered the contractors' questions and exchanged ideas on improving the draft. By the time the final RFP was released on 18 May, little misunderstanding about what the Air Force intended by the stipulations in the document was possible.

The RFP stated that contract award would be based on ''an integrated assessment of offeror's capability and capacity.'' In descending order of importance the major assessment areas were overall engine capability, readiness and support, life cycle cost over a 20-year period, and program adequacy and competition. The RFP further stated, ''Consideration will be given to the effects of dual awards on acquisition and ownership costs, system readiness and availability, and the industrial mobilization base.'' Several unique Air Force requirements in the RFP reflect both a strong desire for contracting flexibility and a forceful reaction to controversy over spare parts acquisition policies and warranties.

The Air Force asked contractors to submit proposals covering an estimated 2,000 engines for the F–15s and F–16s spanning fiscal years 1985–90. Asking for firm prices for each of several different contracting alternatives, the Air Force intended to select only one at the time of award. First, the contractors were asked to offer their best prices if the Air Force chose to buy initially only the fiscal year 1985 requirements with priced options for 1986 and 1987 and not-to-exceed prices for the three years after that. Second, contractors were also asked for their best prices if the Air Force committed to a three- or five-year multi-year contract. Presumably, prices for the multi-year contract would be lower because of scale. However, a multi-year commitment would cost the Air Force the flexibility in future years to switch contractors based on changing needs of the Service. The Air Force might want to keep uncertainty and competition working by making annual buys through successive years to keep the contractors on their toes. Third, all proposals were to include not-to-exceed prices for the Navy, if the Navy were also to buy the selected engine.

From the outset, the Air Force emphasized it had no preconceived ideas about either awarding all the requirements to just one contractor or splitting the requirements between in a particular ratio. The final decision would be made by the Secretary of the Air Force based on a careful analysis of all the contractors' offers. As part of the process, though, contractors were also asked to submit their prices based on split award assumptions (75–25, 50–50, and 25–75) as well as a complete (100 percent) award.

During the summer and fall of 1983 the imbroglio over abuses in spare parts pricing intensified greatly and appeared often in daily news coverage. A provoked Secretary of Defense Caspar Weinberger said, "The system needs fundamental change. These are terrible contracts. We are just not going to pay these prices anymore."[43] Earlier, a proposed "Defense Audit Service Report on the Procurement of Aircraft Engine Spare Parts" draft "had been made public by the Project on Military Procurement, a nonprofit organization ... [with] a network of informers in the Department of Defense."[44] The report included horror stories of grossly overpriced engine parts purchased by the military directly from Pratt and Whitney. For instance, a disk estimated at $1,736.20 wound up costing the Air Force $9,325.78; a $12.00 tube, $639.29; and a $6,445.65 case assembly, $45,236.11.[45]

Competition, instilling an important corrective measure into the spare parts acquisition process, worked well in one case history. The audit cited Pratt's offer to sell the Air Force 15,658 "divergent nozzle assemblies" for $2,469 each. B.H. Aircraft of Farmingdale, New York, submitted an unsolicited proposal to sell the same assemblies for $1,395 each. When not approved as a source for certain technical reasons, B.H. Aircraft protested that decision and eventually received an order for 7,000 of the parts. The Government saved almost $9 million through competition on just this one purchase.[46]

Secretary of the Air Force Verne Orr, a take charge administrator and savvy businessman, studied the problem and insisted the RFP for the fighter engines require usable reprocurement data and dual sources of supply. Seeking assurance that the Air Force would not be locked into future expensive sole-source contracts with either Pratt or GE, Secretary Orr directed his Deputy Assistant Secretary Lloyd Mosemann to ensure that each contractor provide engineering reprocurement data with unlimited rights to the Government. Any limited rights still claimed by the contractor had to be identified and option prices for the Government provided. Moreover, each contractor had to warranty that his reprocurement data were accurate, complete, and adequate for their intended purpose.[47] The Air Force was determined to get the quality of data needed at the time needed in order to buy spare parts at competitive prices. Subsequent changes to the RFP incorporated all of these stipulations. Another novel feature of the RFP was the warranty provision, also a reflection of the times. Seeking a form of insurance against faulty products, Senator Mark Andrews from North Dakota successfully sponsored legislation requiring the military to obtain the same kind of protection against faulty performance as, for example, a farmer would expect in buying a tractor. Warranty protection made a hot topic in procurement circles, and the fiscal year 1983 DOD Appropriation Act included specific direction pertaining to the engine competition:

> None of the funds made available in the Act . . . shall be made available for the purchase of the alternate or new model fighter aircraft engine that does not have a written warranty or guarantee attesting that it will perform not less than 3,000 tactical cycles.

Mr. Daniel Rak, Assistant General Counsel of the Air Force, believes the Andrews warranty law gave the Air Force leverage to negotiate for the kind of warranty the program office had wanted even prior to the Andrews legislation.[48] Therefore, as a consequence of law, coupled with independent Air Force commitment to shift responsibility for the cost of performance failure to the contractor, the RFP included unprecedented warranty provisions.

The RFP's warranty required three basic areas of coverage. First, for three years or 1,000 flight hours, the engine would be free from defects in material or workmanship. If defects were discovered, the contractor would fix them or pay the Air Force to do so. Second, the high-pressure turbine and combustor were covered for eight years or 3,000 tactical cycles. As its objective, the Air Force wanted coverage for a long-term durability, the primary driver of depot overhaul workload. If an engine failed to retain 98 percent of its thrust or exceeded 105 percent of specified fuel consumption, the contractor was required to repair or replace it. Third, in order to prompt the contractor to design a reliable and durable product for the long term, the contractor had to guarantee a ceiling for engine removal rates from 1989 to 1995. The Air Force wanted to ease the workload on maintenance technicians. Obviously, the Air Force constructed each of these exacting requirements primarily to preclude anything even close to the nightmares experienced in maintaining the F100.

THE UNSOLICITED PROPOSAL AND OTHER TACTICS

While the Air Force was drafting and refining the RFP, particularly regarding the pricing matrix, reprocurement data, and warranty provisions, Pratt actively sought derailment of the competition in a variety of ways. In a prime example, an unsolicited proposal Pratt offered to the Air Force in early December 1982 proposed a multi-year, fixed-price contract to deliver almost 2,300 F100s with a series of interesting quality and cost warranties and guarantees. Pratt believed the proposal would save the Government about $3 billion, but the Air Force's reception was cool.

Although Pratt offered a good proposal, the Air Force believed an even better deal awaited by continuing with the competition.[49] The Air Force well understood that the unsolicited proposal was a Pratt effort to circumvent competition. Pratt was obviously worried enough to offer the Air Force what the company had refused to give in 1977 and 1978.[50] Commercial sales had not materialized as Pratt had expected, and now GE was making a strong run at Pratt's once secure military engine business. If just the prospect of competition could force Pratt to offer the improved pricing and warranties, the Air Force believed unleashing the full force of competition through the RFP might bring even better concessions. Furthermore, in studying Pratt's offer carefully the Air Force found some of the alleged benefits tainted. General Skantze, Commander of Air Force Systems Command and another major advocate of competition, stated:

> If you read the fine print, what it really said was, "If you put up the money to support completion of the DEEC development, and the improved life core, and the gear pump, we'll turn around and offer your warranty after those things are completed. . . . *If they were really convinced they could do that, there wouldn't have been the catches.*[51]

Pratt knew that the DOD acquisition community closest to the situation was impressed by the unsolicited proposal, though not enough so to cancel or even delay the competition. To impress people even more, Harry Gray took Pratt's offer directly to the senior leadership in Congress, the President's staff, the Office of the Secretary of Defense, and the Air Force. General Kelly Burke, Air Force Deputy Chief of Staff for Research and Development, said, "I think [Gray's] letter went to every four star, and some three stars in the Air Force. All this was most unusual, to go offline, outside the source selection process like that. It was unprecedented in my experience."[52]

Working the Washington scene rigorously was UTC's Clark MacGregor, formerly a Congressman and Richard Nixon's presidential campaign director. Among those contacted in the Executive Office of the President were James Baker, Edward Harper, and David Stockman. However, what may now be the best known lobbying attempt of all is Gray's letter in January 1983 to Caspar Weinberger. It became known as the "Dear Cap" letter and read, in part,

> Cap, if the quests for defense savings are real ... then it's important this offer (the unsolicited proposal) be recognized.... The Air Force's plan to re-compete the procurement of F–15 and F–16 power plants as now programmed can do nothing but raise acquisition costs.... Suggest a rescheduling of the engine competition to a later model aircraft.[53]

All over Washington UTC conducted a full-court press to convince the nation's leadership that Pratt's unsolicited proposal was an unbeatable bargain, that continued pursuit of the competition was excessively wasteful and, moreover, wasn't fair to Pratt. On the Hill, congressional hearings on the competition provoked an interesting exchange between Representative Patricia Schroeder and Dr. Tom Cooper. Representative Schroeder expressed concern about politicizing the source selection process. Dr. Cooper responded with a slide showing the organization of the source selection process, involving several hundred people. He attempted to stress the numerous checks and balances as well as the number of people involved precluding political control of the process. Congressman Schroeder interrupted him and said:

> We've seen those panels over again and you know my comment—which is sexist and I shouldn't say it—but that is, if those guys were women they'd all be pregnant because they can't say "no" to anything, and we've had panels for everything and they say "yes" to everything.[54]

UTC relentlessly pressed its case, but in mid-February the Air Force told Pratt that "in view of the upcoming fighter competition and source selection, further consideration of your offer is inappropriate."[55] The Air Force believed it could do better. Furthermore, UTC had not favorably impressed the Air Force acquisition community with its unprecedented end run tactics. Neither incentive nor desire was sufficient to alter the Air Force source selection plan. Moreover, the Air Force could not technically accept the proposal. At least, that is the fine point on which the Air Force rejected Pratt's offer. According to General Skantze, it "was submitted when the draft RFP was already out for competition, so in accordance with our defense acquisition regulations, we could not entertain it and had to consider it an early proposal."[56]

Pratt's reaction to the Air Force rejection was mixed, but primarily the company had no alternative to competition. If Pratt wanted to stay in the fighter engine market, and it certainly did, it had to play by Air Force rules. Late in February, after receiving the Air Force letter of rejection, Pratt executives met for seven hours to reassess company posture. After the meeting, Carlson immediately wrote a letter to company employees, declaring a positive attitude about the competition with GE. The letter read, in part, "Let's stop wringing our hands, we haven't lost and don't intend to, so let's tell the world that.''[57] However, while Carlson was pumping sunshine to his staff, he still seemed to harbor some ill will over his perception of broken promises and certain restrictive procurements by the Air Force. During an interview with a local newspaper reporter, Carlson vented his feelings:

> [Regarding] the development of two new technology busting commercial engines plus a third engine being developed . . . Pratt's biggest concern right now is to get those three engines into production in time for an expected revival of the commercial market. If the company succeeds, *it won't mean a tinker's goddamn whether the F100 stays or goes, not a damn thing.*[58]

Although Pratt had agreed to participate in the competition, there were continuing hints it would not do so gracefully. Misunderstandings and failed reconciliations had festered too long on both sides. Pratt never totally ceased its efforts at least to delay the competition in hopes its more powerful engine could demonstrate superior capabilities and compete with the higher thrust GE engine (eventually called the F110). For example, Congressman Addabbo tasked the Surveys and Investigations (S&I) Staff of the House Appropriations Committee to review the competition, including the logistical implications of dual sources. The resultant S&I report generally criticized the Air Force plan because the F100 was now greatly improved, dual sources would create very expensive logistical burdens, and the Soviet threat now required something more powerful than either of the competing engines. Unsurprisingly, the Air Force disagreed with the S&I assessment, and Secretary Orr explained to Congressman Addabbo that the competition would continue. Of course, the Secretary's commitment did not settle the issue on the Hill; problems with Congress continued right through the announcement of the award and explanation of the decision. Nevertheless, despite its frequent inquiries and expressions of concern, Congress did not interrupt the source selection, except to require that the Secretary of Defense himself review and approve the selection decision. Nor did anyone within the Reagan administration, but outside the DOD, intercede forcefully on Pratt's behalf.

However, there were delays in the source selection caused by some in the Office of the Secretary of Defense. One office in particular strongly insisted that the competition be changed to focus on getting higher thrust engines. Another branch of OSD challenged the schedule for timely completion of key testing requirements. All of this rancor spilled out of the Pentagon, and the spillover effects renewed interest and questions in Congress. Still, in spite of tussle after tussle over almost every conceivable aspect of the competition, the source selection continued. Secretary Orr, General Skantze, Dr. Tom Cooper, and a core of other determined stalwarts fought off every charge. They were determined to prove that competition could solve the engine woes as no other alternative could.

THE AIR FORCE'S DECISION

Best and Final Offers arrived on 5 December. Evaluation took just over a month before the Secretary of the Air Force heard the results. Under Air Force acquisition regulations, the Secretary was not given a specific recommendation. The award decision on who would get what was truly his. Meanwhile, in the final days of deliberations, Pratt continued to pressure for support of its product through appeals to the public. A full-page advertisement in the *Washington Post* on 1 February proclaimed the success of a newer Pratt engine in completing a demonstration test requirement. Again though, for the Air

Force, this approach of going directly to the public only served to deteriorate, not in the slightest way enhance, Pratt's image.

Finally, on 3 February 1984, after Secretary Weinberger had approved Secretary Orr's decision and after the Congress had been notified, the Air Force announced a split award. Only the fiscal year 1985 quantity of 160 engines was awarded, and of that, 120 (75 percent) were contracted with GE for F–16s and the remaining 40 (25 percent) with Pratt for F–15s. Secretary Orr, the Source Selection Authority, had decided to wait and observe contractor performance and field experience for a year. Thanks to the pricing matrix completed in the contractors' proposals, he continued to have flexibility each year in selecting his preferred approach.

The Air Force press release on the day of the announcement explained that GE had offered lower overall support costs, had ensured better procurement of spare parts through an outstanding plan for second sourcing and reprocurement of engine components, and had offered an excellent warranty. Basically, GE had been more responsive to the Air Force requirements, and the company's larger share of the first-year buy reflects that fact. In congressional hearings shortly after the awards were made, Dr. Cooper and Colonel Nelson expanded on the companies' proposals and the Air Force evaluations. One key distinction emerged immediately: Pratt had an inflexible, take-it-or-leave-it approach in its proposal.

In what most observers consider an audacious act, which predictably backfired, Pratt discouraged awarding it anything less than 100 percent of the requirements. Dr. Cooper and Colonel Nelson reported that in the Pratt proposal:

> For anything less than a 100% buy, there is a step function that goes to over three times the cost that we expect the engine to cost us in the field.... If we buy one engine less than 100%, our cost goes from being about 9% to being about three times what we think our actual cost is going to be.[59]

In contrast to this Pratt proposal, GE's warranty costs actually decreased as their share of a dual award declined.[60]

Pratt told GAO their warranty terms reflected their corporate strategy to accept greater technical and cost risks for producing all 2,000 engines and less risk on a dual award basis.[61] The consequence of that pricing strategy, however, almost eliminated Pratt from any portion of the fiscal year 1985 contract requirements. Secretary Weinberger and Secretary Orr had to be convinced that giving Pratt at least some share of the initial award was important for continuing to foster competition for the engines. Pratt almost got nothing.

Even now it's not easy to understand Pratt's proposal. The signals from the Air Force should have been clear to Pratt. The Air Force intended to get the benefits of competition and retain them for years to come—that would have been unlikely if the Air Force accepted an all-or-nothing proposal from Pratt. The Air Force had wanted all the blocks in the pricing matrix completed with competitive offers, but Pratt's offers for anything less than 100 percent of the requirements were deliberately noncompetitive. Pratt seemingly tried to intimidate the Air Force into staying with the company as the single source of supply. However, the days of intimidation were past; the Air Force was not in the slightest way inclined to tolerate disregard for the Service's long-sought-for and carefully structured competition.

On the bright side, both contractors' proposals brought substantial benefits in engine operability, supportability, and retention of performance. The new engines were warranted to be twice as durable as the then current F100, free from rapid-throttle-movement worries and from afterburner flameouts, and almost completely relieved from needing extensive ground trimming. In addition, support costs were expected to decrease by approximately 50 percent.[62] Furthermore, GE's proposal had received outstanding ratings for its plan for open competition in the purchase of spare parts for their engine, the F110. General Electric even went beyond the unprecedented requirements in the RFP for establishing dual sources and provided an innovative plan that carried vendors, selected by GE or the Air Force, up to first piece qualification.[63]

General Electric pledged not only to find and train dual sources, but also to stay with them until the Air Force

The General Electric F110 engine.
(Photograph courtesy of Headquarters, US Air Force.)

was satisfied that second sources had really been established. Again, GE skillfully "played" to the Air Force's prime interests. Certainly at that period of time especially, creative techniques to lower the cost of buying spare parts meant a great deal to the military. All the while that Pratt seemed deliberately set on irritating the Air Force, GE continued at every opportunity to give the Air Force at least what it asked for and sometimes even more. General Electric was determined to succeed.

It is difficult for a corporate outsider to know for certain why Pratt structured its proposal as it did. Facts are unavailable, but persistent rumors say that Harry Gray, against the advice of most, if not all, of his staff, shaped the final form of the Pratt proposal. Even if he did not, the proposal at least had his blessing because Gray is well-known for his involvement in the detailed workings of his corporation. As pointed out in a *Business Week* article on Gray's corporate expansion efforts, "Unlike most other takeover artists, Gray is personally involved in the most detailed aspects of his operations.... He even rewrites or edits UTC's advertising.... Gray is criticized for his insistence on remaining so intimately involved in operating details and on running what appears to be a one-man show."[64] With prudent conviction, it's safe to attribute the proposal tactics to Harry Gray.

Remember Pratt's perspective. In the early 1970s, the company had a solid lock on the fighter engine business. The problems with the F100 were not Pratt's alone; responsibility was shared, perhaps evenly, with the military. Pratt believed that the Air Force made promises it didn't keep. Moreover, Pratt felt excluded from competing for some new programs, and Pratt still believed the Air Force was foolish to exclude more thrust as a requirement in the competition. Pratt's dealings with the Air Force had become highly personalized. Strong feelings existed on both sides. As one congressional aide knowledgeable of the history of the competition commented, "Long ago it went well beyond arguments about technology to a political football."[65] Pratt's excellent proposal, satisfying all requirements but discouraging a split purchase, reflects an emotional flavor of the competition as well as the company's hard-nosed business decision to pressure the Air Force to award them the entire buy.

REACTIONS

Unanimous applause greeted the Air Force for its source-selection decision. Frequent critics of military procurement practices praised this Air Force acquisition. Senator Dodd from Connecticut commended Secretary Orr for the cost savings achieved in the split contracts.[66] The major national newspapers, such as the *Wall Street Journal* and *New York Times*, carried lengthy, generally positive stories of the award. One leading aerospace executive said, "This shows that, to survive in this business, you must be just as diligent in producing reliable products as in making your sales pitch in Washington."[67] A Wall Street executive quipped, "This shows there is nothing sacred in dealing with the Government. The Government is becoming as tough as the airlines."[68] The enthusiastic, supportive comments, of course, came as welcome tonic to the Air Force acquisition community, which felt it had been doing a good job all along but had been misunderstood in many of its programs. Even Pratt's supporters in Congress were, in effect, praising through their muted responses.

The Assistant Secretary of the Air Force, Dr. Tom Cooper, explained the rationale for the selection decision to anyone in Congress interested in hearing. Many were. He traveled wherever it was necessary, carrying the message. After Cooper reached a Connecticut group, the *Manchester Journal Inquirer,* the local newspaper to many Pratt employees, printed an article describing reactions, "State Congressional Delegation Backs Off Pact Attack."[69] The temperate views of the delegation, after hearing Dr. Cooper's presentation, are perhaps best summarized by Representative Gejdensen's comment, "If the underlying question is 'Did we find a smoking gun that indicated the contract was unfair?' We did not." Dr. Cooper had been persuasive, and the facts spoke for themselves. Pratt had made an excellent offer for getting everything, yet GE's offer for 100 percent was also competitive. However, Pratt's bids for a split award were not close, and GE won cleanly.

The General Accounting Office (GAO) just about settled any lingering skepticism of the Air Force's performance. Senators Chiles (Florida) and Weicker (Connecticut) had asked the GAO to review the engine source selection, and in late April the GAO replied, "We have concluded the Air Force acted properly and reasonably in making this award." [70] The GAO report noted that the Chairman of the Cost Analysis Improvement Group on Secretary Weinberger's staff described the Air Force's methodology in developing life cycle costs as "the most comprehensive and thorough analysis I have seen."[71] GAO recognized that the most critical part of the decision was splitting the award; therefore, it scrutinized that aspect with particular care. Nevertheless, the GAO found no reason to criticize Secretary Orr's decision, stating, "In our opinion the Secretary's reasons [protecting against work disruptions, increasing contractor responsiveness, and expanding the mobilization base] . . . provide sound bases at this time for splitting the 1985 award and continuing the competition."[72] The Air Force had so carefully and prudently conducted the source selection that no one could find fault. To the contrary, most observers found much to praise.

THE WATERSHED EFFECT

A few days after announcement of the Air Force decision, Secretary of the Navy John Lehman told the Senate Armed Services Committee that the Navy had chosen GE's F110 to replace Pratt's TF30 in the F–14. F–14 crewmen had been dissatisfied for years with the TF30 throttle restrictions and the limitations, caused by the engine, on aircraft performance. The Navy, of course, had had a close association with the history of the F100 and the F110. The Air Force and Navy started the F100 program together. The Navy's withdrawal from the program in the early 1970s created a severe financial problem for the Air Force that, consequently, may have constrained improvements to the engine during development. On the other hand, the Navy's cooperation in transferring most of the $41 million to the Air Force enabled continuing development of the F101X technology at a critical time when otherwise GE might have stopped work altogether.[73] Furthermore, most people familiar with the history of the F100 do not fault the Navy for having pulled out in the 1970s. The Navy wanted a higher thrust engine, tried unsuccessfully to get the fan to work satisfactorily, and felt they received less attention than they wanted from the Air Force.[74]

During the competition, Dr. Cooper kept his counterpart in the Navy, Dr. Paisley, generally informed on the progress.[75] The Navy also stayed fairly close to progress in the competition at the program office. However, the consensus is that the Navy would have selected the F110 for the F–14 regardless of the Air Force decision.[76] This does not mean the Navy would have selected the F110 without the Air Force participation in the program. With huge bills for building a 600-ship Navy, the Navy needed the Air Force flight data, competitive prices, and economies of scale. The Navy needed a new engine for the F–14 but, as in the past, could not afford a stand alone retrofit program. As explained by Secretary Lehman, "The only reason we can afford . . . [the engine] now is that the Air Force is developing it for the B–1B bomber and developing a lighter version for its own fighter. We are simply parasiting on their having paid for the bill through the B–1 and their own RDT&E program."[77] The Navy's $33 million "investment" in the Air Force a few years before, though at the instigation of the Air Force and with crucial congressional support, was paying off well and in so doing was fulfilling what Congress had originally intended for the money anyway, a new engine for the F–14.

All through the competition, Pratt was mindful of what effect the Air Force award decision might have on foreign sales of the F100, a lucrative market. Not only did most foreign air forces want the same equipment flown by the US Air Force, but also the attraction of a more powerful alternative to the F100 would probably be irresistible. Unsurprisingly, in May 1984 Israel selected the F110 to power an additional 75 F–16s being ordered from General Dynamics. Similarly, Turkey contracted for 173 F110s for their F–16s. The loss to Pratt was more than just splitting the fiscal year 1985 US Air Force requirements. Pratt began losing overseas sales which would have been theirs, if there had not been the competition.

Interestingly, however, many observers in the United States and overseas failed to realize that, even with the 75–25 split in the first year award, Pratt still received more of the Air Force's total fiscal year 1985 business than did GE. Because the source selection was completed more than a third of the way into the fiscal year, the Air Force had to protect aircraft production schedules and had contracted earlier with Pratt for a separate quantity of 120 engines on a sole-source basis. Adding these 120 to the calculation of the fiscal year business base, the total business for the year split at approximately 60 percent to Pratt and 40 percent to General Electric.[78] However, this reality paled in the hoopla over Pratt's loss of any part of the business.

With the source selection completed, competition continued for the following years of military requirements, and the power balance between GE and Pratt shifted dramatically toward GE. Were the results of the competition worth the mighty struggle?

6
OVERALL ASSESSMENT

On the previous page, the beneficiaries and victors in the ''Engine War,'' F–16 Falcon aircraft, shown in an air-to-air right side view as one breaks to the left. (An official US Air Force photo, Ken Hackman, photographer.)

THE ENGINE COMPETITION is foremost a tribute to the strong character of a small core group of public servants, scattered at different levels throughout the Government, all similarly bound by an aggressive determination to get a better engine and responsive service. Very few of those involved actually believed the GE engine would be purchased in large numbers, if at all. Through the first steps in evaluating proposals, many insiders still felt the competition little more than an elaborate hoax to spur Pratt to respond meaningfully to Air Force concerns. Amazingly, at competition's end GE received a contract for 75 percent of the first year requirements. Fourteen years for Pratt. Thousands of engines to power F–15s. Yet, in unprecedented fashion, the Air Force re-introduced competition.

Vision and perseverance by a committed few are primarily responsible for this competition achievement and the initial benefits to the Government. No one could have anticipated or arranged the fortuitous confluence of events that made the competition possible. The believers, the few who saw competition as the way to get a better engine, had the tenacity to keep the idea moving forward, around, and through a maze of obstacles.

The operational benefits from the F110 and the improved F100 are impressive. They provide, for example, unrestricted throttle movement, improved war-time surge capability, improved thrust, and lower fuel consumption. So far, the Air Force has been unable to induce an F110 stall stagnation under any condition and expects close to the same results for the F100 with the DEEC. Dollar-wise, the Air Force expects to reduce the cost per flight-hour, maintenance man-hours, and engine removals per 1,000 flight-hours by 30–50 percent.[1] Furthermore, the warranty provisions for cost and risk sharing, plus the ability to compete the procurement of spare parts, came through the competition.

However, it's early to make a final evaluation of the competition now. Although early evidence strongly suggests the competition was completely successful, final evaluation will depend on operational experience with the engines and on how prudently the Air Force takes advantage of the many opportunities in the contracts. What is already absolutely clear, though, is the importance of maintaining the competitive environment as far into the future as possible.

Although the Air Force has the option to select a single contractor instead of splitting future requirements between two, doing so would be a mistake. Having labored so valiantly to get two qualified sources, the Air Force must ensure that competition continues. Already, some Air Force insiders worry that too much military business belongs to GE, perhaps jeopardizing Pratt's future in the engine business and promoting a GE arrogance to customer needs. However, continuing competition assures marketplace forces work on behalf of Air Force interests.

F100 DEVELOPMENT

Between 1970 and 1980 at least 16 studies, assessments, and investigations discussed the development and acquisitions of engines.[2] A number of these reports found fault with the way the F100 was developed. Operational problems with the engine are attributed to management decisions by the SPO. However, when Air Force management receives careful consideration in the complete context of that time, vigorous criticism of SPO decisions seems unjustified. Similarly, Pratt cannot be faulted too severely for its engine design work. The Air Force asked for a race horse, and Pratt produced one. Durability, reliability, operability, and maintainability were not specified contractually. Now, with our greater understanding, for example, of duty cycles and thermal fatigue, such criteria certainly would be.

All available literature and all authorities on jet engine development strongly advocate more time for research and testing. For example, the Air Force Scientific Advisory Board points out, "Development of a new engine is to a large degree an empirical process requiring many 'builds' and tests before satisfactory design is achieved."[3] Some experts on engines believe designs require three to four more years to complete than the airframe. Yet, schedule pressures typically force concurrent engine and airframe development, with the airframe setting the obligatory pace.

Why can't you start engine development well ahead of airframe development so that engine capabilities can significantly affect aircraft design? In fact, programs exist today to get a head start in developing advanced propulsion technologies. However, in the late 1960s, at go-ahead time for the F–15, the Air Force had to move forward quickly with the best available technology. The F–15 had to be operational by November 1974. The foreign threat, Navy ambitions, congressional monitoring, Air Force pride—these were the realities and drove early completion. No one doubts more time should have been allowed ideally for F100 development. However, realities do not always accommodate the ideal; they certainly didn't for this program.

When the F–15 and F100 programs started, the Soviets threatened to control the skies with their advanced fighter aircraft. Also, the Air Force had recently been stung by harsh charges of mismanaging the C–5 acquisition. In addition, the Air Force had already piggybacked on two successful Navy fighter development programs (F–4 and A–7) and did not want to do so again. However, the Navy had ambitions to bail out the Air Force once more, this time with the F–14. At the same time, the Office of the Secretary of Defense was intent on proving to the public that DOD could do business smartly, and the Air Force top leadership was especially keen to meet flawlessly all program milestones. Such were the realities. Never mind the ideal. The Air Force had to produce. Reasonable people had to make difficult decisions quickly.

One man, charged by the Deputy Secretary of Defense with responsibility for success or failure of its program, directed both the airframe and engine development. Because of the risks in this combined organizational scheme, we haven't used the approach again. However, in 1969, the objective was to get "rubber on the ramp" by November 1974, and single-manager accountability seemed the sound way to facilitate the process. Though shortcuts might be necessary to meet the objective, with Pratt's cooperation and some additional money, anything could be fixed, or so some may have thought. Any delay in the program schedule put the F–15 at risk of cancellation—an intolerable alternative. Moreover, General Bellis believes no major propulsion advances have been made since the F–15 and, if significant advances are needed, a combined airframe and engine program arrangement will be essential. The general states, "With the Air Force's new Advanced Technology Fighter (ATF) program, I would assume there will be a 'single' arrangement approach or I can assure you the propulsion system of inlet engine, augmentor, and thrust vector devices will never be successfully developed."[4]

The most frequently cited suggestion of management indiscretion in the early F100 program is General Bellis' decision to waive a portion of the Military Qualification Test (150-hour endurance test) to approve the engine for production. His aim was clear: to avoid costly disruption of the aircraft production schedule. And certainly the waiver neither caused nor precluded earlier detection of problems with stall stagnations or turbine failures. However, while General Bellis had the authority to grant the waiver, he should have immediately told his superiors of his action. Once the modified test had been completed successfully, the Government had to accept engines for delivery,[5] and the program schedule was no longer in jeopardy. Bellis' mistake was managerial. Superiors could have, and should have, been fully informed of the MQT decision. However, the entire issue too often gets more attention than it deserves. Its relevance was inflated at the time by those whose self-interests would be served by focusing uncomplimentary attention on the Air Force programs.

What really matters regarding the MQT waiver, and all other management actions accelerating the development process, is what the primary motivation was in pressing forward so urgently for an operational capability by November 1974. If countering the Soviet threat of air superiority in the mid-1970s was truly the main consideration, then the national interest was well-served by compressing the schedule. The performance capabilities of the F–15 are unexcelled, and the performance of the F100 in large measure accounts for those capabilities. Yet, if competition with the Navy and other bureaucratic infighting were the primary motivators, then rigid adherence to schedule was terribly wasteful.

My research led me to conclude that the Soviet threat was the primary motivator of the competition. That threat provided impetus to get the F–15 program started, pushed the program to completion, and produced an aircraft with superb capabilities. The F–15 has one of the best safety records in the history of fighter aircraft.[6] The problems following initial operation stemmed primarily from initially inadequate technical specifications. These inadequacies have been corrected, but for various reasons the Air Force and Pratt could not make the corrections smoothly.

MID AND LATE 1970s

Three principal elements dominated the events of this period and, taken together, threatened national security: technical problems with the engine, inadequate logistical provisioning by the Air Force, and corporate recalcitrance.

Technical problems can be expected in any program which harnesses advanced technology to expand the state of the art. Compared to its predecessor, the F100 generated much more power, with less weight. It used new materials in complex ways never before attempted. Thus, although the stall stagnations and turbine failures were certainly regrettable, failures of some sort were almost inevitable, particularly with a schedule imperative. None of the failures was caused by inattention or sloppiness. They stemmed from pursuing a monumental task, a task technicians were still trying to understand at the time, and from coping with the permutations possible when thousands of interrelated parts combine in newly conceived designs.

No one doubts that technical problems surfaced early after first operational use of the engine. But the problems surfaced, paradoxically, because the engine was such a wonderful initial success. Pilots made their aircrafts perform in extraordinary fashion, stressing engines to unanticipated excess. "Rapid throttle movements, which produced little response from other engines, in the F100 resulted in new combat maneuvers. They also resulted in new problems. Stall stagnation ... and [TAC] cycles now standard measures of a fighter engine's toughness, were both terms invented for the F100."[7] Also, at about the same time, Tactical Air Command inaugurated the Red Flag exercises to emphasize realistic weapon system and pilot stresses in training for air to air combat. This more rigorous peacetime flying helped, inadvertently, to uncover reliability and maintainability problems with the engine earlier through accelerated accumulation of duty cycles.

Unfortunately during this period, the Air Force also made unwise logistical decisions to support the engine. Based, perhaps, on overly optimistic descriptions of engine capabilities and constrained by budgets, the Air Force made decisions in 1975 for small quantities of spare engines, too small to have been prudent even with information available then. In 1975, Pratt was already projecting that 30–50 percent of critical turbine parts would wear out and require replacement at the 900-cycle inspection point. At the rate the engine was being flown and cycles accumulated, 900 cycles would come far earlier than initially projected. The Air Force should have recognized these facts and then ordered and expedited increased quantities of spare parts. Instead of looking at current experience and obvious trends developing through use of the F100, the Air Force was, instead, estimating its logistical needs based on experience with less sophisticated engines in less demanding operations.

However, what aggravated the Air Force the most materially and emotionally was the perception that Pratt and its parent, UTC, were indifferent and arrogant. The ''Great Engine War'' probably would not have occurred if UTC had been oriented toward genuinely assisting the Air Force to resolve tough safety, technical, and management problems with the F100. If the Air Force leadership had felt that UTC was making real efforts to respond to customer concerns, the Air Force probably would have continued to rely on that sole contractor. Sensing that Pratt was interested only in highly legalistic interpretations of contract provisions and in maximizing its profits soured the Air Force. Even later, after management realignments had improved Pratt's image, suspicion and ill humor lingered. Perhaps emotions on all sides had become so piqued that even innocent actions and bland communications became misinterpreted and resented.

For example, during recovery from the Ladish and Fafnir strikes, some Air Force people believed Pratt was cheating on agreements for allocating available output. Some observers believed Pratt was manufacturing and shipping more parts to the commercial airlines and fewer to the Air Force than had been agreed. This assessment, never substantiated, then or since, added fury to the word-of-mouth denigration of Pratt. A senior Air Force officer, familiar with details of the strike recovery, denies any basis for the rumor. He even doubts that Pratt had sufficient manufacturing control to fine-tune material allocation had it wanted to favor the airlines. Nevertheless, this rumor, and others, circulated and heightened tensions. Certainly, mutual trust between the Government and contractors is important for a successful program. When trust erodes, suspicion and miscommunication ensue on both sides.

Pratt did make a fundamental blunder in its dealings with the Air Force. Customer satisfaction fell to a low priority. The incessant pursuit of cash to enrich corporate coffers and thereby facilitate expansion through expensive acquisitions ascended. Pratt ignored a lesson important for anyone involved with the Government contracts. The Government can react shrewdly and relentlessly to pursue its own self-interest in the marketplace. Not bound by past behavior and business relationships, when a contractor appears to take unfair advantage of its contractual relationship, the Government has the responsibility to react creatively. In this case, it did so. In testimony before the House Appropriations Subcommittee on Defense in 1979, General Lew Allen, Air Force Chief of Staff, explained:

> We are *concerned about the motivation and incentive of Pratt and Whitney to correct this engine*. Their full dedication is needed to correct the engine and insure its availability for the very long lifetime expected and because of the enormous reliance that United States tactical forces are placing on the engine. . . . [The] best way to insure that we were adequately addressing the problem was to generate some competition. . . . The approach with General Electric . . . is an attempt . . . to develop a true competitive situation within the engine industry.[8]

The Air Force believed Pratt was not exerting its best effort to improve the engine. Pratt was receiving hundreds of millions of dollars through CIP and spare parts sales. The Air Force saw no financial incentive for Pratt to improve, unless forced by competition.

EARLY 1980s

Wanting to have a competition and having the wherewithal to make it happen were far apart. The biggest issue was money, not money to develop just a new GE engine for demonstration, but to afford the high cost of developing a second engine qualified for production and to create the logistical base to support it in the field.

When the Reagan administration resurrected the B–1 program, the Alternative Fighter Engine program no longer had to bear the full cost of establishing a logistical base for GE's F110. Common parts, common maintenance equipment, common training, and common manuals resulted in a highly beneficial form of symbiosis in which the shared costs for a common engine would help the B–1 program stay within the $20.5 billion program ceiling established by President Reagan and would save the AFE program millions. According to Colonel Jim Nelson, without these savings the AFE would not have been competitive.[9] In addition to the B–1, the re-engined KC–135 was funded to install the fuel efficient CFM56. This engine also was closely related to the F110 and similarly helped reduce F110 production costs.

Commonality provided not only substantial savings for logistical support; it also considerably reduced other front-end costs. For example, the need to design and buy unique tooling was virtually eliminated. GE estimated tooling commonality could save 50 percent or more of comparable costs of starting from scratch. In addition, the initial F110 unit price would be lower because of existing experience in manufacturing the common components used in the F101 family. Consequently, about a 25 percent savings in starting the program was expected in initial production.[10]

The Air Force could not have predicted reviving the B–1 program when the first thoughts of fostering competition arose. Nonetheless, if those few who saw the problem and understood the need for an alternative had not persisted in their beliefs in spite of the many objections, the Air Force would not have been as ready as it was to capitalize on fortuitous events emerging over time.

Good fortune came in surprising forms. Bad publicity about Pratt's spare parts prices embarrassed many Government officials and undermined some efforts to expand the DOD budget. However, the timing of the publicity helped thwart UTC's intense lobbying efforts to postpone, if not cancel, the competition. UTC made strong presentations throughout the Washington power structure. It might have succeeded in its quest had it not been for disclosure in the press of Pratt's alleged overpricing of spare parts. When it came time for a public figure to step in and disrupt the source selection process, no one wanted to champion the Pratt cause. What public figure would stake his reputation on Pratt and against competition when the press was suggesting that Pratt overcharged for parts in a sole-source environment?

To many Government officials, Pratt could not be trusted. Why protect such a company from competition? That question could not be answered in Pratt's favor. In spite of much work to cultivate support for the Pratt position within the Washington community, UTC could not overcome the momentum for competition and the antipathy for Pratt, strengthened by the spare parts controversy.

Again, the core group of competition devotees could not have anticipated the coincidental release of publicity to counter the contractor's lobbying campaign. However, undaunted in its pursuit of what it felt was right, the group was ready when external events, which it did not control, fell into place at crucial moments.

SOURCE SELECTION

The source selection process was conducted faultlessly, not just with respect to the technicalities of the evaluation process. The GAO reviewed the complete source selection record and reported back to Pratt's advocates in the Senate that no impropriety nor unreasonableness in the selection was found. Moreover, the Air Force always dealt with the process and other public inquiries with great finesse.

The Air Force Secretariat carefully ensured that the Secretary, the source selection authority, always made public statements. Air Force officers, with few exceptions, did not make statements on the Air Force position. Every effort avoided suggesting an Air Force general officer versus Pratt emotional battle. Such an appearance might have swung sentiment from the Air Force to Pratt.

PEOPLE AND COMMITMENT

Another fortuitous circumstance was the pattern of Government promotions and transfers to increasingly powerful positions that provided uncommon longevity for key people working this issue.

For all the criticisms military personnel systems receive at times, the relatively frequent reassignments gain merit when, in cases such as these, moves allow dedicated people to keep working important issues from different vantage points. Although the reassignments were not intended specifically to continue work on F100 problems, in most cases, the reassignments reflect the valid rationale that people who had expertise and commitment could be best used in their new positions across a range of related issues. It was not happenstance that people with records of accomplishment continued to work on the same issues.

FUTURE CONCERNS

The key to future evaluations of the competition may be how well the Air Force takes advantage of the warranty provisions and the reprocurement data. The Air Force should capitalize on the unprecedented opportunities offered by these provisions. Unfortunately, budget limitations, complexity of institutional procedures, and overemphasis on expediency may squelch chances to make the competition the great success it could be.

For instance, the Air Force must have enough training and organization to protect its rights under the warranty provisions. Commercial airlines view warranties seriously, often organizing separate offices with the single responsibility of monitoring all repair work to ensure that all airline's rights under warranties are protected. Sometimes airlines contract with companies that provide warranty monitoring service.

However, the Air Force has no comparable organization. And an organizational scheme short of a full-time office, similar to the airline arrangement, might not suffice. In the field especially, the Air Force is dedicated to keeping its aircraft flying rather than determining contractor liabilities and protecting Government contractual rights. But warranties *are* very expensive. If manpower ceilings, budget constraints and workload priorities will not support effective management of warranties, then perhaps the House Armed Services Committee was correct, when after reviewing engine warranties, some members recommended the Air Force buy the engines uninsured.[11] After working with the warranty coverage for a few years, the Air Force should re-examine exactly what the benefits have been.

The situation with reprocurement data is basically the same. Can the Air Force use the data properly? The Air Force struggles even now to keep up with the volume of procurement parts needed to maintain operational readiness. What will happen to the reams of additional reprocurement data when the Air Force gets them? Will the Air Force have the people with the talents to analyze the data and buy parts competitively? Previously that function was performed by the prime contractor. Some people doubt that the Air Force can adjust personnel ceilings and internal procedures to absorb the new workload. Here, too, the Air Force should reassess in a few years the actual benefits from the contract provision.

Whether these provisions were mistakenly structured and incorporated in the contracts, the provisions made sense. Congress even dictated part of the warranty. There was no alternative. The real concern is whether the Air Force has sufficient numbers of knowledgeable, adaptable, success-oriented people to see the potential and use it. Moreover, the skepticism focuses on the Air Force system itself. The Air Force is not bound to act in the ways of the past; the competition proves business does not have to be as usual. However, will the top-level attention and supporting staff tenacity be there to follow through on these new opportunities?

LESSONS LEARNED

The value of starting engine development well in advance of airframe development, whenever possible, and of designing greater durability into the engines in all cases—these are two important lessons learned. However, the most important lesson to draw from the engine experience is the value of competition. Competition is the only sure way to get the best effort.

The $15 million "seed money" Congress authorized in 1976 for a new engine in the F–14 fostered competition in the development efforts. The authorization encouraged GE because it supported an earlier corporate decision to explore fighter applications.[12] Competition, the chance to prove itself and take business away from its rival, motivated GE to invest corporate money on the project and to work aggressively to demonstrate its superior engine capabilities. GE, which in 1975 had no high-thrust fighter engine business, found competition an opportunity, a carrot to chase after.

Over the months of informal and formal competition, GE appeared anxious to please, to satisfy customer concerns. General Electric worked quickly to prove the qualities of its engine. The success of the F101 DFEs during ground and flight tests in the F-16 and F–14 further developed the competitive environment.[13] GE's innovation and risk taking, evidenced by its provisioning and warranty proposals, astutely played to Air Force desires. That responsiveness was appropriately rewarded. Without competition, these benefits would not have been.

Competition to Pratt was a threat, a stick. Anything less than winning 100 percent of the fiscal year 1985 quantities was defeat since no competition meant all the business. Pratt's reaction to the competition was much like its reaction to other Air Force efforts to get an improved engine—Pratt fought the idea. However, competition did yield from Pratt some substantial initial benefits to the Air Force. For instance, Pratt offered engine improvements to the Air Force earlier than the Air Force had been led to expect without the competition. Furthermore, the unit prices were lower than Pratt had previously been offering. Since the initial split in February 1984, competition has further induced Pratt to grant even more concessions to the Air Force. Warranty prices have been reduced significantly and arrangements with the EPG have improved. Competition has extracted from Pratt what trust, personal cajoling, and public rebukes failed to accomplish.

A corollary to the competition lesson is to carry competition as far forward in acquisition programs as possible. Continue the economic rivalry to satisfy military requirements. Realizing this lesson, the Air Force only initially awarded the fiscal year 1985 quantities. This decision seemed particularly wise just months after the award. In the fall of 1984, an Air Force audit team found problems with purchasing systems, quality assurance, and work measurement at GE's Evendale plant.[14] In addition, some insiders worry that in 1986 and later GE might be manufacturing as many as 10 different types of military engines but Pratt only one. Has the pendulum of military business swung too far over to GE? Some who now work daily with GE engineers and lower level managers detect a more hard-nosed, bureaucratic control over the flow of information between the GE plant and the Air Force similar to the 1970s experience with Pratt when communications grew strained and led to much misunderstanding.

The audit team's findings are serious but correctable, not enormously troublesome. The problems stemmed more from complacency and sloppiness during relatively slow periods of business activity than from flagrant disregard of the rules. General Electric's greater business base now is the just reward for successfully competing in the past. The stricter, formalized communications now being imposed are a natural consequence of production expansion and the need to control the quality of administration, in much the same way the Air Force audit team has enforced the contract subtleties more strictly. Nevertheless, by late 1984, regardless of reasonable explanations, the Air Force was wary of the balance of business between GE and Pratt, particularly after Pratt had significantly improved its proposal for fiscal year 1986. Therefore, in January 1985, the Air Force, not unexpectedly in these circumstances, awarded 54 percent of the fiscal year 1986 quantities to GE and 46 percent to Pratt, a clear signal to each to stay sharp. Competition is now a carrot for Pratt and a stick for GE.

Maintaining competition in this way is not always possible and may have limitations even for the fighter engines in future years. To continue to see it work, the Air Force needs high total volume over the entire program and a high enough annual production rate.[15] In explaining the engine competition to the House Armed Services Committee, Dr. Tom Cooper said, ''On this particular program [Alternate Fighter Engine] we are looking at 3,000 engines . . . trying to get dual sources on 50 C–5 engines probably doesn't make much sense.''[16] The volume over the entire program and the annual production rate must be high enough to justify the fixed costs of supporting two manufacturers. The exact breakeven point, where the cost of one versus two sources is equivalent, will vary depending on the specific programs and contractors involved.

Some analysts even question whether the Air Force will ever realize the anticipated $3 billion savings on fighter engines. They point out the estimate assumes the Air Force will purchase 250 aircraft per year, yet the present budget estimate calls for fewer than 160 per year. Without a major increase in defense spending, they argue, the total six-year purchase will be closer to 1,400, rather than 3,000. And this volume might be insufficient to support both Pratt and GE.[17] Therefore, each contractor, knowing this prospect, has extra incentive to perform at its best, though economics may eventually force the Air Force to go 100 percent with one contractor.

The big lesson from the Great Engine War is the value of talented, dedicated people. The successes of the competition are primarily attributable to noble public servants. Press coverage of military procurement leaves many citizens angry over accusations of grossly inept Government business practices. Evidence of exorbitant prices for simple parts and equipment, overruns, and expense account abuses undercut public confidence in Government and tarnish the self-esteem of military and civil service employees. However, in the midst of these ill feelings, the engine competition is a rallying point. The competition proved that the Government can play a high stakes game successfully. Unintimidated, undaunted smart people can solve or avoid the most desperate of problems. The key ingredients for success are committed leadership, resourceful staff, and tenacious workers in the trenches. Everyone who contributed to the fruition of the competition should feel justly proud.

ENDNOTES

PREFACE

1. Interview with Mr. Gene Bryant, Dayton, Ohio, 6 November 1984.
2. Letter from Gen. Benjamin N. Bellis, 18 July 1985.

INTRODUCTION

1. ''Air Force Announces Alternate Fighter Engine,'' Office of Assistant Secretary of Defense (Public Affairs), News Release No. 49–84, 3 February 1984, p. 2.

2. ''Larry Skantze: Instigator of the Great Engine War of 1983,'' *Defense Week,* 29 August 1983, p. 14.

3. News Release No. 49–84, p. 2.

4. John M. Barry, ''How GE Beat UT for Huge Engine Contract,'' *Dun's Business Month,* April 1984, p. 14.

CHAPTER 1

1. "The Dogfight over the F–15," *Business Week*, 20 December 1969, p. 96.

2. Donald C. Winston, "New MiG Imperils U.S. Superiority," *Aviation Week and Space Technology*, 3 June 1968, p. 22.

3. *Jane's All the World's Aircraft*, 1982–83, p. 208; US, Department of Defense, *Soviet Military Power* (Washington, DC: GPO, April 1984), p. 55.

4. "Soviets May be Shifting Design Emphasis," *Aviation Week and Space Technology*, 5 August 1968, p. 33.

5. "New MiG Imperils U.S. Superiority."

6. "FX Proposal Requests Imminent," *Aviation Week and Space Technology*, 2 September 1968, p. 16.

7. Jacob Neufeld, *The F–15 Eagle Origins and Development 1964–1972*, Office of Air Force History, November 1974, p. 25.

8. Neufeld, p. 15.

9. US, Congress, House, Committee on Appropriations, *Department of Defense Appropriations for 1969*, Hearings, 90th Cong., 2d sess., 26 February 1968, p. 706.

10. "Soviets May be Shifting Design Emphasis," p. 33.

11. "FX Proposal Requests Imminent," p. 17.

12. James Fallows, "Trouble with the Air Force's Eagle," *Atlantic Monthly*, January 1982, p. 64.

13. US, Congress, Senate, Committee on Armed Services, *Fiscal Year 1972 Authorization for Military Procurement, Research and Development, Construction and Real Estate Acquisition for the Safeguard ABM and Reserve Strengths*, Hearings, 92d Cong., 1st sess., 8 March 1971, p. 3806.

14. Interview with Lt. Gen. Benjamin N. Bellis (Ret.), Bethesda, Maryland, 27 September 1984.

15. Neufeld, p. 41.

16. "Joint Management Established for F100," *Aviation Week and Space Technology*, 26 June 1972, p. 83.

17. US, Congress, Senate, Committee on Armed Services, *Review of TAC AIR Programs for Fiscal Year 1972 Authorization*, Hearings, 92d Cong., 1st sess., 8 March 1971, p. 3819.

18. Neufeld, p. 55.

19. Ibid.

CHAPTER 2

1. "Joint Management Established for F100," *Aviation Week and Space Technology,* 26 June 1972, p. 83.

2. Interview with Col. James Nelson (Ret.), Dayton, Ohio, 7 November 1984.

3. Edward C. Westwood, "The F–15 Engine Development," Professional Staff Study, Air War College, Maxwell Air Force Base, Alabama, April 1975, p. 22.

4. Interview with Lt.Gen. Benjamin N. Bellis (Ret.), Washington, DC, and Bethesda, Maryland, 27 September 1984.

5. US, Congress, House, Committee on Appropriations, *Department of Defense Appropriations for 1973,* Hearings, 92d Cong., 2d sess., 18 April 1972, p. 1173.

6. US, Congress, Senate, Committee on Armed Services, *Review of TAC AIR Programs for Fiscal Year 1972 Authorization,* Hearings, 92d Cong., 1st sess., 28 April 1971, p. 4024.

7. Ibid., p. 4029.

8. Jacob Neufeld, *The F–15 Origins and Development 1964–1972,* Office of Air Force History, November 1974.

9. US, Congress, Senate, Committee on Armed Services, *Military Procurement, Fiscal Year 1974,* Hearings, 93d Cong., 1st sess., 15 May 1973, p. 1459.

10. Interview with Col. Ray Moss (Ret.), Washington, DC, 17 October 1984.

11. Memorandum from Deputy Secretary of Defense to Secretary of the Navy, 6 July 1971.

12. US, Congress, Senate, Committee on Armed Services, *Fiscal Year 1974 Authorization for Military Procurement, Research and Development, Construction,* Hearings, 93d Cong., 1st sess., 18 May 1973, p. 1667.

13. Interview with Mr. Edward Trusela, Washington, DC, 1 August 1984.

14. House Appropriations Committee, 18 April 1972, p. 1175.

15. Westwood, p. 69.

16. US, Congress, Senate, Committee on Armed Services, *F–15 and F–16 Engine Problems,* Hearings, 96th Cong., 1st sess., 27 November 1979, p. 3.

17. House Appropriations Committee, 18 April 1972, p. 1172.

18. Westwood, p. 77.

19. US, Congress, Senate, Committee on Armed Services, *Military Procurement, Fiscal Year 1974,* Hearings, 93d Cong., 1st sess., 26 April 1973, p. 1117.

20. General Bellis, interview.

21. Letter from General Bellis, 18 July 1985.

22. US, Congress, House, Committee on Armed Services, *Cost Escalation, Indefinite Procurement Contracts and Military Posture,* Hearings, 93d Cong., 1st sess., 22 May 1973, p. 1389.

23. Westwood, p. 34.

24. US, Congress, Senate, Committee on Armed Services, *Department of Defense Authorization for Appropriations for Fiscal Year 1980,* Hearings, 96th Cong., 1st sess., 2 April 1979, p. 1216.

25. Interview with Mr. Harvey Gordon, Bethesda, Maryland, 5 October 1984.

26. General Bellis, interview.

27. Senate Armed Services Committee, 2 April 1979, p. 1231.

28. Ibid., p. 1202.

29. Westwood, p. 30, table 2.

30. US, Congress, House, Committee on Armed Services, *Military Power,* Hearings, 96th Cong., 2d sess., February 1980, p. 2266.

CHAPTER 3

1. Comptroller General. *Report to the Congress of the United States on Status of the Air Force's F–16 Aircraft Program,* 24 April 1978, p. 8.

2. Ibid., p. 12.

3. Maj. William C. Morrison, "How Safe is the F–100 Engine?," *Aerospace Safety,* August 1979, p. 9.

4. Joel Lang, "The Great Engine War," *Hartford Courant* (Northeast Magazine), 27 March 1983, p. 12.

5. Interview with Gen. William Evans (Ret.), a UTC executive. Washington, DC, 21 November 1984.

6. Interview with Gen. Alton Slay (Ret.), McLean, Virginia, 24 November 1984.

7. Interview with Gen. Richard Steere, Andrews Air Force Base, 27 November 1984.

8. Interview with Col. James Nelson (Ret.), Dayton, Ohio, 7 November 1984.

9. US, Congress, Senate, Committee on Armed Forces, *F–15 and F–16 Engine Problems,* Hearings, 96th Cong., 1st sess. 27 November 1979, p. 13.

10. US, Congress, Senate, Committee on Armed Forces, *Department of Defense Authorization for Fiscal Year 1980,* Hearings, 96th Cong., 1st sess., 2 April 1979, p. 1208.

11. Comptroller General Report, 24 April 1978, p. 9.

12. US, Congress, House, Committee on Appropriations, *Department of Defense Appropriations for 1980,* Hearings, 96th Cong., 1st sess., 15 February 1979, p. 467.

13. US, Congress, House, Committee on Appropriations, *Department of Defense Appropriations for 1980,* Hearings, 96th Cong., 2d sess., 20 February 1980, p. 303.

14. Interview with Lt. Col. Phillip Hughes, (Ret.) Evendale, Ohio, 8 November 1984.

15. Ibid.

16. US, Congress, Senate, Committee on Armed Services. *Department of Defense Authorization for Appropriations for Fiscal Year 1980,* 96th Cong., 1st sess., 2 April 1979, p. 1199.

17. Interview with Gen. Alton Slay (Ret.).

18. US, Congress, House, Committee on Armed Services, *Department of Defense Authorization of Appropriation for Fiscal Year 1984,* 98th Cong., 1st sess., 27 April 1983, p. 1319.

19. Ibid., p. 1322.

20. Senate Armed Services Committee, 27 November 1979, p. 19.

21. Interview with Mr. William Cowie, Dayton, Ohio, 7 November 1984.

22. Interview with Mr. Joseph Wood, Dayton, Ohio, 7 November 1984.

23. USAF Scientific Advisory Board, *Special Report on Turbine Engines,* January 1976, p. 3.

24. Morrison, p. 9.

25. Comptroller General Report, 24 April 1978, p. 9.

26. Senate Armed Services Committee, 27 November 1979, p. 26.

27. Morrison, p. 9.

28. Senate Armed Services Committee, 2 April 1979, p. 1224.

29. Senate Armed Services Committee, 27 November 1979, p. 28.

30. Ibid., p. 30.

31. Ibid., p. 4.

32. "Contention Over F100, F101X Intensifies," *Aviation Week and Space Technology,* 31 March 1980, p. 79.

33. House Appropriations Committee, 20 February 1980, p. 306.

34. Comptroller General Report, 24 April 1978, p. 11.

35. Senate Armed Services Committee, 27 November 1979, p. 24.

36. Ibid., p. 16.

CHAPTER 4

1. Barry Bluestone, Peter Jordan, and Mark Sullivan, *Aircraft Industry Dynamics* (Boston: Auburn House Publishing Company, 1981), p. 20.

2. Ibid., p. 23.

3. Ibid., p. 39.

4. Ronald Fernandez, *Excess Profits* (Reading, Massachusetts: Addison-Wesley Publishing Company, 1983), p. 220.

5. Ibid., p. 224.

6. Ibid.

7. Ibid., p. 237.

8. *Business Week,* 10 December 1979, p. 77.

9. Ibid., p. 74.

10. *Forbes,* 15 April 1975, pp. 42 and 44.

11. General Electric Co., *Seven Decades of Progress* (Fallbrook, California: Aero Publishers, Inc., 1979), p. 12.

12. Ibid., p. 39.

13. Ibid., p. 51.

14. Bluestone, p. 63.

15. *Forbes,* 18 August 1980, p. 35.

16. Bluestone, p. 62.

17. Ibid., p. 64.

18. General Electric, *Seven Decades,* p. 147.

19. Gerhard Neumann, *Herman the German* (New York: William Morrow and Company, 1984), p. 99.

20. Ibid., p. 236.

21. Ibid., p. 247.

22. *Headlines,* General Electric, Evendale, Ohio, 4 October 1976, p. 1.

23. Interview with Mr. Morris Zipkin, West Palm Beach, Florida, 12 November 1984.

24. Interview with Mr. Nicholas Constantine, Alexandria, Virginia, December 1984.

25. Testimony of Honorable Thomas Cooper before the HASC on 27 April 1983, p. 1317.

26. John M. Barry, ''How GE Beat UT for Huge Engine Contract,'' *Dun's Business Month,* April 1984, pp. 40–44.

27. Interview with Col. Kenneth Eichmann, December 1984.

28. Testimony of Gen. Alton Slay before the SASC on 27 November 1979, p. 49.

CHAPTER 5

1. Interview with Mr. Joseph Wood, Dayton, Ohio, 7 November 1984.

2. Interview with Lt. Col. Phillip Hughes (Ret.), Evendale, Ohio, 8 November 1984.

3. Ibid.

4. Interview with Mr. Nicholas Constantine, Alexandria, Virginia, October 1984.

5. Interview with Gen. Alton Slay, McLean, Virginia, 24 November 1984.

6. Ibid.

7. Ibid.

8. Ibid.

9. US, Congress, Senate, Committee on Armed Services, *F–15 and F–16 Engine Problems,* Hearings, 96th Cong., 1st sess., 27 November 1979, p. 65.

10. Harry Gray, letter to General Allen on 2 February 1979.

11. US, Congress, Senate, Committee on Armed Services, *Department of Defense Authorization for Fiscal Year 1980,* Hearings, 96th Cong., 1st sess., 2 April 1979, p. 1227.

12. General Slay, letter to General Dixon, dated 15 March 1979.

13. John M. Barry, "How GE Beat UT For Huge Engine Contract," *Dun's Business Review,* April 1984, p. 41.

14. Interview with General William Evans (Ret.), a UTC executive, Washington, DC, 21 November 1984.

15. Robert Carlson letter to Dr. De Lauer, Under Secretary of Defense for Research and Engineering, dated 4 August 1982.

16. US, Congress, House, Committee on Appropriations, *Department of Defense Appropriations for 1980,* Hearings, 96th Cong., 1st sess., 15 February 1979, p. 462.

17. Ibid., p. 471.

18. Senator Weicker letter to Dr. Perry, dated 6 February 1979.

19. House Appropriations Committee hearings on 15 February 1979, p. 463. (Emphasis added.)

20. Ibid., p. 471. (Emphasis added.)

21. US, Congress, Senate, Committee on Armed Services, *Department of Defense Authorization for Appropriations for Fiscal Year 1980,* 96th Cong., 1st sess., 2 April 1979, pp. 1245–1246. (Emphasis added.)

22. Barry, p. 42.

23. Interview with General Slay.

24. Interview with Mr. James Day, Dayton, Ohio, 7 November 1984.

25. Senate, Committee on Armed Services, hearings on 27 November 1979, p. 95.

26. Interview with Mr. Morris Zipkin, West Palm Beach, Florida, 12 November 1984.

27. Winston Williams, "Engine Maker Works on Image," *New York Times,* 26 December 1979, pp. D1–D2.

28. Interview with General Slay.

29. General Poe letter to General Allen, dated 29 Septamber 1980.

30. Senator Chiles letter to Senator Stennis, dated 17 October 1980.

31. General Allen letter to Senator Stennis, dated 18 November 1980. (Emphasis added.)

32. Congressman Addabbo letter to Dr. Mark, dated 19 December 1980.

33. Brigadier General Murray letter to Congressman Addabbo, dated 13 January 1981.

34. Interview with General Steere, Andrews Air Force Base, 27 November 1984.

35. Interview with Mr. Zipkin.

36. Interview with Mr. Constantine.

37. Interview, Colonel James Nelson (Ret.), Dayton, Ohio, 7 November 1984.

38. Interview with Mr. Dean Gissendanner, Washington, DC, 30 October 1984.

39. Interview with Colonel Nelson.

40. Walter S. Mossberg and Edward T. Pound, "How Pratt and Whitney Gains from the Way US Buys Spare Parts," *Wall Street Journal,* 3 October 1983, p. 25.

41. Ibid., p. 1.

42. Ibid.

43. Ibid.

44. Charles Mohr, "Pentagon Audit Finds Sharp Price Increase for Parts," *New York Times,* 23 July 1983, p. A13.

45. Draft Proposed Defense Audit Services Report, "Procurement of Aircraft Engine Spare Parts" (3AP–021), 10 June 1983, p. 29.

46. "Pentagon Audit Finds Sharp Price Increase for Parts."

47. US, Congress, House, Committee on Armed Services, *Alternate Fighter Engine,* Hearings, 98th Cong., 2d sess., 8 March 1984, p. 10 (transcript).

48. Interview with Mr. Daniel Rak, Washington, DC, 27 July 1984.

49. Interview with Mr. Gary Hansman, Dayton, Ohio, 7 November 1984.

50. Interview with Colonel Nelson.

51. "Larry Skantze: Instigator of the Great Engine War of 1983," *Defense Week,* 29 August 1983, p. 15. (Emphasis added.)

52. Barry, p. 43.

53. Harry Gray letter to Secretary of Defense Weinberger, dated 31 January 1983.

54. US, Congress, House, Committee on Armed Services. *Defense Department Authorization and Oversight,* Hearings, 98th Cong., 1st sess., 27 April 1983, p. 1346.

55. Lt. Gen. Thomas H. McMullen letter to Mr. Frank W. McAbee, Jr., (President of Pratt), dated 14 February 1983.

56. House Armed Services Committee hearings, 12 April 1983, p. 906.

57. Joel Lang, "The Great Engine War," *Hartford Courant* (Northeast Magazine), 27 March 1983, p. 19.

58. Ibid., p. 12. (Emphasis added.)

59. House Armed Services Committee hearings on 8 March 1984, p. 48.

60. Comptroller General, *Review of the Alternate Fighter Engine Competition (GAO/NSIAD–84–104),* 17 April 1984, p. 7.

61. Ibid.

62. John B. Nix and Riley C. Shelnutt, "Behind the Alternate Fighter Competition," *Aerospace America,* May 1984, p. 73.

63. House Armed Services Committee hearings on 8 March 1984, p. 53.

64. "What makes Harry Gray Run?," *Business Week,* 10 December 1979, pp. 77 and 80.

65. Raymond Banner, "10 Billion Jet Engine War," *New York Times,* 14 January 1984, p. L27.

66. "Roth Praises Air Force for Jet Engine Contracts," *Defense Daily,* 24 February 1984, p. 307.

67. Richard Witkin, "Decision Big Setback for Pratt" *New York Times,* 4 February 1984, p. 38.

68. Ibid.

69. Miriam Silver, "State Congressional Delegation Backs off Engine Pact Attack," *Manchester Journal Inquirer,* 16 February 1984.

70. Comptroller General, 27 April 1984, p. 1.

71. Ibid., p. 7.

72. Ibid., p. 2.

73. Interview with Mr. Constantine.

74. Interview with Mr. Gissendanner.

75. Interview with Dr. Thomas Cooper, Washington, DC, 24 November 1984.

76. Ibid.

77. "Navy Studies Retrofitting F–14 With F110 Engines," *Aviation Week and Space Technology,* 16 July 1984, p. 26.

78. House Armed Services Committee, hearings on 8 March 1984, p. 13.

CHAPTER 6

1. US, Congress, House, Committee on Armed Services, *Air Force Alternative Fighter Engine, Hearings before the Subcommittee on Procurement and Military Nuclear Systems,* 98th Cong., 2d sess., 8 March 1984, p. 30 (transcript).

2. Comptroller General, *Report to Congress of the United States, Are Management Problems in the Acquisition of Aircraft Gas Turbine Engines Being Corrected?* 30 September 1980, p. 61.

3. USAF Scientific Advisory Board, *Report of the Ad Hoc Committee on Engine Development,* August 1973, p. 2.

4. Letter from Gen. Benjamin N. Bellis, 18 July 1985.

5. US, Congress, Senate, Committee on Armed Forces, *Fiscal Year 1974 Authorization for Military Procurement Research and Development, Construction Authorization for the Safeguard ABM, and Active Duty and Selected Reserve Strengths,* Hearings, 93d Cong., 1st sess., 18 May 1973, p. 1652.

6. Henry L. Littlejohn, "Today's Jet Engines Are Better Than Ever," *Flying Safety,* April 1984, p. 5.

7. Joel Lang, "The Great Engine War," *The Hartford Courant* (Northeast Magazine), 27 March 1983, p. 13.

8. US, Congress, House, Committee on Appropriations, *Department of Defense Appropriations for 1980 Hearings before a Subcommittee of the Committee on Appropriations,* 96th Cong., 1st sess., 15 February 1979, p. 390. (Emphasis added.)

9. Interview with Col. James Nelson, USAF (Ret), Dayton, Ohio, 7 November 1984.

10. James N. Krebs, written statement prepared for the House Armed Services Committee, May 1983.

11. "Skip Engine Warranties, Air Force Is Advised," *Washington Post,* 11 May 1984, p. A21.

12. "The F100: An Important Program." *Headlines,* General Electric, Evendale, Ohio, 4 October 1976, p. 1.

13. John B. Nix and Riley C. Shelnutt, "Behind the Alternate Fighter Engine Competition," *Aerospace America,* May 1984, p. 72.

14. "Air Force seeks F100 Warranty Price Cut," *Aviation Week and Space Technology,* 22 October 1984, p. 27.

15. *Hearings before the Subcommittee on Procurement and Military Nuclear Systems,* 8 March 1984, pp. 68 and 69.

16. US, Congress, House, Committee on Armed Services, *Department of Defense Authorization of Appropriation for Fiscal Year 1984,* 98th Cong., 1st sess., 27 April 1983, p. 1329.

17. Richard Barnard, "GE Wins AF Engine Race," *Defense Week,* 3 February 1984, p. 1.

INDEX

B

THE AUTHOR

COLONEL ROBERT W. DREWES, US Air Force, wrote this book while attending the Industrial College of the Armed Forces as a National Defense University Senior Fellow. He has over 15 years of contracting experience, including assignments at the Aeronautical Systems Division, Air Staff, and the Office of the Assistant Secretary of the Air Force for Research, Development, and Logistics. He has also served on the staff of the Executive Office of the President. Colonel Drewes earned his BA in economics from Colby College and an MBA from the Harvard Business School. He is a distinguished graduate of the Air Command and Staff College and a graduate of the Air War College.

THE AIR FORCE AND THE
GREAT ENGINE WAR

*Composed in 11 on 13 point Times Roman
with 30 point bold chapter titles.
Cover text in ITC Benguiat Italic and Zapf Light.
Mechanicals by Laszlo Bodrogi and Nancy Glover.*

*Editorial Reader: Captain Herbert Smith,
English Department, US Air Force Academy.
Design and Edit: Janis Bren Hietala
Editorial Assistant: Pat Williams*